The Lincolnshire Poachers

10 . 11 . 01 .

good health, good fortune
and happy reading

Derek Mills.

Mr David Nickerson presenting the author with a clock to mark his retirement.

THE
LINCOLNSHIRE
POACHERS

Derek Mills

ASHRIDGE PRESS

Published by:
Ashridge Press
20 Oderin Drive, Trafalgar Park, New Waltham, near Grimsby DN36 4GJ

ISBN 1 901214 80 X

By the same author:
A COVEY OF TALES:
THE MEMOIRS OF A LINCOLNSHIRE GAMEKEEPER

FOREWORD

When I read the stories in this book my mind went back to the many nights Derek spent out in all weathers waiting for poachers.

After, sometimes hours, of worrying, it would be such a relief to have him return to bed (with hands and feet like blocks of ice) but at least he was safe and well.

His seasonal vigils obviously had some reward because his estate encountered few poaching incidents and none of them, thankfully, involved violence.

After 45 years he has put all that behind him and written this book to recount the many poaching tales he has remembered which have happened to him or his keepering friends.

Although the word 'poaching' paints a picture of romantic roamings in the countryside, when reading these tales, spare a thought for the keeper's wife and family and the hours of worry some of them have endured, all in the name of game protection.

Gill Mills

THE LINCOLNSHIRE POACHER

When I was bound apprentice in famous Lincolnshire,
Full well I served my master for more than seven year
Till I took up to poaching as you will quickily hear;
O 'tis my delight on a shining night, in the season of the year.

As me and my companions were setting of a snare,
'Twas there we spied the game-keeper, for him we did not care,
For we can wrestle and fight, my boys, and jump o'er anywhere;
O 'tis my delight on a shining night, in the season of the year.

As me and my companions were setting four or five,
And taking on 'em up again, we caught a hare alive,
We took the hare alive, my boys, and thro' the woods did steer,
O 'tis my delight on a shining night, in the season of the year.

I threw him on my shouldier, and then we trudged home,
We took him to a neighbour's house, and sold him for a crown,
We sold him for a crown, my boys, but I did not tell you where,
O 'tis my delight on a shining night, in the season of the year.

Success to every gentleman that lives in Lincolnshire,
Success to every poacher that wants to sell a hare,
Bad luck to every game-keeper that will not sell his deer,
O 'tis my delight on a shining night, in the season of the year.

(Traditional)

INTRODUCTION

The opening line of Percy Grainger's famous song 'The Lincolnshire Poacher' goes: 'When I was bound apprentice in famous Lincolnshire…'

The scenario he had in mind when he wrote what has become this county's anthem was the widespread practice at the time of young lads being articled as apprentices to different tradesmen at pitifully small wages. Their income was relied upon to help with the general household expenses and, to help makes ends meet, they simply turned to poaching.

These were the true Lincolnshire Poachers, the men who took what Nature provided to help make a living and to put food on the family table. They operated mainly in their own locality but, as time moved on and transport became more widely available, the poacher himself became more mobile and began to cover an ever-widening area.

As farming methods changed natural wild game became less plentiful and more game was hand-reared and this, as we shall see later, led to more extensive poaching and far bigger bags. In turn this led to poaching becoming attractive to criminal gangs and, as a result, to a serious increase in rural crime before it was made less profitable by the fall in game prices.

With this background in mind, I have endeavoured to explain the various and varied poaching incidents which I came across during my long career as a game-keeper, incidents which occurred throughout the large county of Lincolnshire.

The mainstay of most poaching 'bags' over the years has

7

been the pheasant and, as something of an insight into the lifestyle of the wild Lincolnshire pheasant, we begin our story of the county's poachers with a look at the subject from the point of view of the quarry...

A PHEASANT'S TALE

I first saw the light of day one June morning through a mass of grass and leaves as I peeped out from beneath the feathers of my brooding mother before she gently turned and lifted me from our nest of dead grass and broken egg shells.

My siblings and I — there were ten of us altogether — had all hatched the previous evening from our shells, a task achieved with much effort and a lot of pecking and cheeping. After a night's rest, during which the warmth of our mother's body dried our striped and mottled down, we had all regained our strength and were eager to see what lay outside our nest.

We all gave off our plaintiff little cheeps and mother reassured us with a low cluck. We pecked at everything which attracted our attention and attempted to swallow bits of grass and leaves before mother gently lifted us out of the nest and began to walk away. She gave a few clucks and we eagerly scampered after her.

The sun shone down through the long grass and the bank we were on was very warm. After a while and lots more pecking, mother gave a few clucks and we all ran to her. She picked up some insects in her bill and dropped them close to us. We immediately began to peck at them and swallowed a few. They tasted very sweet and were much better than the leaves and grass we had had before.

Now the aphids and other insects on the bank seemed much more attractive to us than the grass and leaves and we soon became very adept at catching them for ourselves.

As it began to get dark mother selected a very sheltered place under a hedge not far from the nest site. We all tucked ourselves into her feathers and that's where we spent the second night of our lives.

The following morning the sun shone as brightly as the day before and the bank along the hedge again became very warm. We all cheeped intermittently but mother reassured us with low soothing clucks to let us know all was well.

It was on this third day that we began to venture a little further into the standing grass and corn. Mother appeared ever watchful, standing with head cocked to one side, surveying the sky above and most of the surrounding area. We chicks busied ourselves, pecking at insects and anything else that caught our eye. There did seem a particular abundance of small green caterpillars of the sawfly. They were particularly tasty and by eating lots of them we all began to grow very quickly.

Within four days we had got several nice mottled feathers along both our wings. They were light fawn with black bars and matched the rest of our bodies and helped us blend in with our surroundings.

Alongside the bank and hedge where we had spent our lives so far ran a small dyke and at its bottom was a trickle of running water. As the sun shone day after day it was thirsty work keeping up with our diet of green caterpillars and wild flowers and we would slide down into the dyke before taking a few sips of the crystal clear water. This helped refresh us and gave us new energy as we carried on foraging for more insects.

All the time mother kept a watchful eye on us. One day she gave a sudden sharp cluck of alarm and we chicks all squatted motionless as a large moving shadow formed on the bank side close by. Sometimes it stopped before moving on, only to stop again. While the shadow was over us not a single chick moved and it was only some time after it had gone that we resumed our foraging for food. When the danger had passed we answered mother's cluck and she told us the shadow had been

cast by a hunting kestrel on the look out for an easy meal and
if we ever encountered anything like it again we were to lie
motionless and hope that our camouflaged bodies would save
us from certain death.

At about this time a large chestnut breasted pheasant with a
broad white ring around its neck and a long flowing tail
arrived to join us. Mother said this was our father but, because
he had another three wives and families, he took only a passing interest in us. She told us that we were all descended from
birds which had arrived in England from across the sea in
Scandinavia and many of us lived as part of the wild bird population in the surrounding area.

Father sometimes joined our brood for a short while before
moving on to oversee his other families. He was a magnificent
specimen, periodically puffing out his chest and calling out
what sounded like 'Cock-up, cock-up' This display was,
apparently, to impress his harem and to ensure any rivals kept
out of his territory.

One day as we were busy looking for food we heard father's
distinctive cry, but this time the tone was lower and we knew
instinctively it signalled danger. We were all on the alert when
suddenly there was a movement in the grass and a brown and

white animal dived out of the grass and grabbed one of our brood. There was a cry of alarm from one of my sisters but it was too late as the strange animal disappeared into the ditch with her grasped tightly in its mouth. Mother told us later she had been taken by something called a stoat and it would be best if we moved from the area in case the stoat came back looking for another meal. So quickly we moved out into the corn where mother said it was much safer than the hedgerow, which was much favoured by stoats for their hunting.

All this time we were growing and more feathers were appearing on our backs and necks as well as on our wings. After about four weeks mostly spent in the corn we had a very heavy rain storm. We got very wet and the corn itself was soaked so, when the rain stopped, we worked our way to the edge of the field until we were out of the corn and where we could dry ourselves in the warm sunshine. But we had not been basking for long before there was a sudden 'whoosh' and another of my sisters was grabbed by a bird we were later told was a sparrow hawk. It struck her with lightning speed, grabbing her in its talons and carrying her off in a flurry of feathers.

The next day we were still in the same area when a similar thing happened. This time it was one of my bothers that was taken, again with the same terrifying suddenness. Mother realised we were very vulnerable and took us all back into the shelter of the corn.

We remained there for another two weeks before we came out again, this time going into the shelter of a large hedgerow. The following morning there was a heavy dew and the grass in the hedgerow was quite damp. By this time our wings had grown quite well and we were all eager to try them out. Mother hopped onto a rail in the hedgerow and we all followed, using our wings to get from the ground to the top of the fence. Using our wings for the first time was a wonderful feeling and it also meant we were out of the damp grass and up in the sunshine where we would dry out quickly.

That evening mother flew from the ground and landed in a small bush in the hedge. She clucked and gradually persuaded us to follow her. This we did and spent the whole night perched in the bush as mother thought we would be safer off the ground because, as we grew bigger, we were in danger of becoming prey to a fox.

After this first night, each time the sun started to go down we looked for a bush in which to roost. We tended to pick the same one and there we would sleep until first light the following day. We all slept close to one another and would each occasionally chirp to each other during the night.

As the corn grew we began to feed more and more on the ripening grain. This went on until the corn completely changed colour and then, one day, there was a terrific noise and a large red machine came charging through the field. Although we were in a family group, the machine came so fast and made such a noise that we were all frightened and ran and flew in all directions. We all escaped except for one of my brothers who, for some inexplicable reason, ran towards the machine and was promptly squashed by one of the large black tyres which carried the machine along. The rest of us escaped to the nearby hedgerow and rejoined other members of our family.

After the machine had finished work in the field it left a short stubble and quite a bit of spilled grain. This came in very useful for a feed and we did not have to travel far before our crops were full. By now we were nearly as big as mother and every evening now we went to our roosting bush.

As the days shortened the nights became colder and gradually the leaves began to fall from the trees and the hedgerow bushes. As we sat on our familiar branches we could now see in the bright moonlight what was going on around us.

One evening after feeding on the stubble all day and eating only the tastiest morsels, we settled down to go to sleep. We hadn't been asleep long when we were awoken by the nearby pigeons clattering from a bush. Then, with our eyes wide

open, we could see the shadowy forms of two humans approaching our bush. I had never been so close to a human before but I thought they could be dangerous so I clattered out of the thorn bush, making a lot of noise in the stillness of the night. The rest of the family followed me and we all flew off in different directions, skimming over the darkened fields before finally landing with a bump hundreds of yards away.

Now safely on the ground I made myself as comfortable as possible until day break. As the sky began to lighten I called in my new cock pheasant voice: 'Cock-up, cock-up'. It was only then that I noticed I had landed in the middle of the same stubble field where we had been feeding recently and, as the light grew stronger, I could see mother and one of my sisters some way away as they started to feed. I walked over to join them and, as I arrived, my brother and two more sisters arrived and we were able to reform our family group. However, my other brother was still missing.

I kept calling but my cry went unanswered and we began to search the field. As we made our way along the hedgerow towards our roosting bush we came upon a large bunch of his feathers and the bloody remains of his legs. Mother said he must have been very unlucky for, after the poachers frightened us out of the bush, he must have been killed and eaten by a fox as he slept on the ground. This just emphasised how important it was to roost in a bush. Even though it was chillier on frosty nights, it was far safer than sleeping on the ground.

We shared the remainder of the harvest with other birds, mainly wood pigeons, but gradually the grain became more difficult to find and we had to scratch amongst the loose straw to find it. We still remained with mother and one day she guided us to a lone tree on the far side of the stubble, an area I had not been to before. The tree was a large one and underneath the ground was covered with acorns. Mother pecked at some and then began to swallow them before we followed her example. At first we were reluctant to swallow such a large seed but, after a few attempts, we managed it and soon began to enjoy this new addition to our diet.

That evening as dusk approached we were still busy gorging ourselves on the acorns when mother suddenly flew up into the branches and we all followed. I landed on a large branch and it certainly made more sense to sleep in this oak tree than fly back across the field to our thorn bush, particularly because the large boughs of the oak gave us some protection from the cold winter winds which were beginning to sweep across our field. The other benefit was that the following morning we could fly down from our roost and start breakfasting on the sweet smelling acorns. I am sure it was the smell of the oak tree which had attracted mother to the tree in the first instance because when we arrived at the oak the previous day the first thing we had noticed was the sweet aroma of ripening acorns.

After a few weeks all the acorns had gone and mother decided it was time to move again. We finally left behind the field which had been our home for so long and made for some distant trees. When we arrived we found a lot of pigeons feeding on a brown carpet of fallen beech nuts. These were small three-cornered seeds which had fallen from the large trees overhead. These beech nuts were particularly tasty and, although we needed a lot more of them than we did acorns, we soon had our crops bulging.

By mid afternoon mother said it was time to be making our way back to the oak tree and we soon followed her, arriving just before dusk. The following morning as it began to get light we flew out of the oak tree and headed back to the beech trees, just beating a flock of pigeons to the beech nuts. A road ran alongside the beech trees and it carried a fair number of cars, which ran over and squashed the beech nuts which had fallen onto its surface. These squashed nuts were even more tasty than the whole ones and before long, while the pigeons fed on what had fallen from the trees, we walked out into the road to peck at the squashed nuts, moving aside to let the occasional car pass.

This pattern of life continued for a few days until the day when one of my sisters ran across the road just as a car approached at a very high speed. She had seen some squashed nuts in the centre of the road and was too busy eating them to notice the car approaching. Suddenly there was a bump and a cloud of feathers and her lifeless body rolled onto the road-side.

After this we were not too keen to feed on the road, preferring the safety of the field for a few more days until all the beech nuts had been eaten. Mother then decided it was time to go in search of another source of food and we travelled from our oak tree along a hedge until we came to a plot of kale.

The kale was quite tall but, apart from the shelter it provided from the wind, I couldn't understand why mother had taken us there. We walked about among the kale, pecking at odd weeds before mother found some fresh grain. We all fed heartily on this new supply of food and, finally satisfied, we made out way back to the oak tree in the late afternoon.

During the night it became very cold and the next morning

the whole of the surrounding fields were covered in a white, icy frost. We were all very hungry and came down from our roost and headed straight for the field of kale. Again there was some fresh wheat scattered in the kale, which was very fortunate because everything else was covered in frost and was uneatable. We fed on the grain and a lot more of my relations — cousins, aunts and uncles — arrived to feed with us.

By the fourth morning there must have been more than fifty of us feeding in the kale, including some birds I had never seen before and it became a race to get there early to get the best pickings. The cock-birds from the group I had not seen before had different colourings from my brother and me; they were very lightly coloured and had a broad white ring on the neck like me but where my back had a green sheen, theirs was pale blue.

They were very aggressive and chased my brother away from the feed and began to peck him. He didn't like this and flew from the kale in the direction of the beech trees and I never saw him again. They then turned their attentions on me but I was made of sterner stuff and stood my ground. I made sure I got to the feed very early and got a good feed and then kept away from the intruders, returning to the grain just at dusk and getting a quick feed before going back to the oak tree.

When mother returned to the oak tree she was upset because my brother had been chased away and, with the other losses to our brood, we were down to just four, my three sisters and me. My sisters didn't seem too bothered about their other brother being chased away as the new arrivals had made quite an impression on them. They told them stories about how they never went to roost in trees but slept on the ground where it was warmer. This impressed my sisters until mother reminded them quite sharply about what happened to our brother who was killed by a fox after we had fled from the thorn bush on the night the poachers came.

Mother told us the new arrivals were of French origin, their ancestors having first arrived in France by way of first China and then America. She said they were well known for wandering about the countryside like avian gypsies, moving from place to place as the whim took them. Hence their arrival in our plot of kale.

The next day dawned dull and windy and we left our oak tree and arrived at the kale and began searching for the grain. We were scurrying around picking up a few bits left from the previous day but, for some reason, there was no new grain that day. Everyone was a bit uneasy, fighting and pecking over every grain found when, suddenly, there was a loud bang. This startled everyone and I made my way to the edge of the kale and looked out. The crop seemed to be surrounded by men and some brown and white animals which I had never seen before.

As the men advanced through the kale some of the birds took to the wing and flew in the direction from where the bang had come. I decided to stay where I was, on the edge of the kale with mother and my sisters while we tried to decide what to do.

Soon after the first birds took to the wing there were a lot more bangs and I could see some of the birds falling from the sky. Just at this point, one of the brown and white animals came rushing through the kale straight for our family. I told them to follow me quickly and we all immediately jumped into the air. Once airborne, I could see a line of men where the bangs were coming from and, as I did, I could see more birds falling from the sky. But there did seem to be a gap where no one was so I headed for it straight away, followed by my mother and sisters. It meant flying into the strong wind which wasn't too much of a problem for me but the four females, who were all lighter than me, were caught by a strong gust and lifted high into the air.

Although I was still fighting the wind, I managed to keep an eye on them and, as they flew over the line of men, I heard ten bangs and saw mother and one of my sisters fall from the sky.

The other two managed to get high enough to escape from danger.

I flew on for some distance, finally landing exhausted in a hedge. Back on the ground I could still hear the bangs in the distance and then it all went quiet. By late afternoon I had made my way back to the kale and there I found plenty of fresh wheat. It hadn't been there in the morning and must have been left by the men before they left. I ate quickly and got a good crop full before heading back to the oak tree, flying to my usual roosting branch with my usual 'Cock-up, cock-up' call and began to settle down to go to sleep. Then there was a flapping sound and my remaining two sisters fluttered up from the ground and joined me.

By this time it was almost dark and, as we huddled together, they told me that after they had flown over the line of men the wind had taken them for about a mile before they had finally landed. They had hidden in a thick hedge until the danger had passed. Then they had walked out into a nearby field where they met our brother. They said he looked well with colourful plumage and they had told him about the ambush in the kale field and the loss of our mother. They tried to persuade him to come back to our roosting tree and described how delicious the wheat in the kale field was and the many relatives he would meet there. But he told them he was quite happy foraging on his own for what food he could find in the hedgerows. The news that my brother was nearby cheered me up somewhat even though we had lost our mother and another of our sisters earlier in the day.

After meeting our brother, the two females had walked against the wind and eventually found the fresh wheat in the kale field. They had also had a quick feed before hurrying back to the oak tree.

Later in the night the moon came out and it became very cold. There was a distant bump and the whole area echoed to the sound of cock pheasants calling after being disturbed from their sleep. I listened as the calls became ever more distant and

felt relieved that one of them might have been made by my brother. It had been a sad day for me when he had been chased away by the blue-backed gypsies.

Time passed and things returned to normal. The days grew shorter and the wheat was scattered in the kale field for us every morning. We ate a good breakfast and then wandered around in the kale for a few hours before eating what remained as the light began to go and headed back to our oak tree.

One morning my sisters and I arrived but we could find no fresh grain. I immediately told my sisters to follow me because I remembered what had happened the last time and I led them into a nearby hedgerow where we pecked at a few bright red berries. We hadn't been there long when we could hear the sounds of another ambush in the kale field. We could hear lots of loud bangs and we stayed where we were until they had all finished. Cautiously we made our way back into the kale and there we found lots of fresh grain. But we were the only birds there, except a few small finches. All the other pheasants and wood pigeons had been frightened away or shot by the men.

The same pattern continued through that cold winter. Each time no new grain arrived my sisters and I went to the hedgerow and avoided the ambush before going back to find fresh food had been put down.

As the days began to lengthen and the sun became a little warmer the grain continued to arrive each day in the kale but the ambushes stopped. Life now became much better for us all as we no longer had to be constantly on the look-out for the men and their dogs. Several of us had survived the shooting season and we mostly continued to live around the kale field, including several of the blue-backed gypsies. One of these was paying particular attention to my sisters, who thought his plumage was wonderful. I didn't agree with their choice and eventually I decided to go back to the field where mother had brought us up. It seemed to me to be an ideal habitat, particularly as there was always fresh water nearby in the dyke.

When I arrived I found the field had been cultivated and sown with a small, round seed I later discovered was called a tic-bean. There seemed to be a lot of these seeds on the ground and I was able to eat my fill before I wandered over to the thorn bush where we had first roosted.

I spent that night alone and the following morning I flew down into the field, landing with my usual 'Cock-up' call, letting all those around know who and where I was. Not too far away I heard a similar call and I knew then I had company though who it was I had no idea.

Later in the morning, after a good breakfast of tic-beans, I wandered in the direction of where I had heard the call come from. I went along the dyke and, on rounding a bend, I came across a blue-backed cock accompanied by four females. He rushed at me in a challenging manner and it was clear he had violence in mind. By this time I had grown two very long spurs on the rear of my legs and they were very sharp making them ideal fighting weapons. I noticed the blue-backed cock also had spurs but mine looked longer and sharper.

He came right up to me and told me I was on his territory and if I didn't go back the way I had come he would fight me. I'd never liked these blue-backed invaders since the day they chased my brother away from the kale field and I was in no mood to be frightened off. I sized him up and then challenged him by running straight at him and jumping with my legs stretched out, slashing at him with the spurs. One of the spurs caught his neck and tore out a few feathers. He immediately retaliated and for a short while spurs and beaks were making feathers fly. Eventually, I got the upper hand and, after one particular telling peck from my sharp beak, he retreated, skulking away into some nearby reeds.

That night as I went to my roost I was joined by two of the females who had been with the blue-back. To tell the truth I wasn't really looking forward to the next morning because I knew Mr. Blue-back was still in the area and I just knew our little battle wasn't yet over. Sure enough, just as dawn was breaking there was a screech and some flapping from the direction of the reeds. It seemed a vixen had grabbed the blue-back and bitten off his head. The two females still with him immediately took flight along the dyke.

I must admit to thinking that the vixen had solved my problem for me. And it served the blue-back right for ignoring all the advice and sleeping on the ground. Although a fox had earlier killed my brother, this seemed to even up the score in my mind.

The remainder of his harem soon joined us so, in the space of a day, I had acquired four wives. We soon mated and they brooded eggs and all the while I kept guard on the area, just as my own father had done, warning them of any danger.

After the chicks were hatched and reared, I went into the corn fields which later, after the harvest, became stubble fields once more, feeding on the left-overs. Then it was time for the acorns, followed by the beech nuts, then the winter fed grain.

And so the cycle went on for fours years. During this time I learned to spot danger and how to avoid it, particularly the kale field ambushes.

As my fifth winter approached I was feeding in the centre of a large stubble field when I heard two large bangs not far away. I looked round and the field seemed to be surrounded by men with those same brown and white dogs. I squatted in the stubble, lying flat to the ground, hoping I had not been seen. I remained there for some time as the men and dogs approached. One of the dogs had obviously caught my scent and came rushing towards me. I realised I had been seen and, after taking a few steps, I quickly became airborne.

Making use of the stiff breeze, I tried to gain height quickly, knowing that the higher I could fly the better my chance of escape was. I soared on the wind, flying directly over one of the men. I saw him raise his gun and saw the puff of smoke as he fired. Immediately by right wing collapsed and I began to plummet towards the ground some distance from the men.

I struck the ground with a thud which initially stunned me. I rolled over, shook myself and, realising my legs were still good, set off at a fast pace for a nearby dyke. I had almost reached it when I was overtaken by a dog which grabbed me and started to carry me back to his owner.

I knew then the game was up and in the few moment left to me I cast my mind back to the wonderful life I had had and how it might never have been had I succumbed to the poachers' bullet on that long-ago November night in my very first year.

THE PARKER BROTHERS

The reputation of the Lincolnshire Poacher is known throughout the world and has probably evolved from stories similar to that you are about to read...

The Lincolnshire Marsh stretches along the coastal plain from south of Cleethorpes to north of Skegness has always been dotted with small villages and isolated farm cottages, most of its inhabitants working and living off the land.

Large areas of the land within the Marsh area was traditionally owned by absentee landlords who would send their cattle there for summer grazing. So, naturally, during the winter months and the game season very few people were about on these small marshland fields, surrounded by high thorn hedges and deep drainage ditches.

It was a habitat which was ideal for wild game — hares, pheasants and partridges — while the drains themselves offered excellent refuges for all kinds of wild fowl. With virtually every field having a different owner and most of them living many miles away, the local pot-hunter (the poacher) had an almost free hand to do as he wished.

Times were very hard in the 1930s when men on the Marsh had mainly large families and very little work to support them. It was against this background that many farm labourers and smallholders turned to shooting to provide meat for their family's table. I have heard it said that times were so bad that on one winter's day in the village of Fulstow (a renowned stronghold of pot-hunters to this very day) when the tracks of a hare were seen in fresh snow, within hours 18 men, all armed, were out hunting for this prize.

Every village on the Marsh could boast several families of pot-hunters. One of the best known of all were the Parker brothers from Alvingham.

Tom, Henry and Bob all owned smallholdings around the village and shot over their own land... and everyone else's land for several miles around.

When I first met Tom he told me of the time when he and Henry spotted a cock pheasant on a snowy winter's day. It was one o'clock in the afternoon and they immediately decided to go out and shoot it. On their approach, however, the pheasant took fright and flew across a couple of fields before alighting in some particularly deep snow.

They carefully followed the pheasant across the fields to where it had first landed and then along a couple of dykes in the direction it had subsequently taken. By this time it had flown into the adjoining parish but they continued to follow its tracks.

They lost the trail of the elusive pheasant at one stage but, on checking back, they managed to find it again and followed the running bird. As they approached it once again took fright and flew back in the direction it had come from. They carefully marked it down into a deep dyke and retraced their steps. Then Henry decided to take a short cut and headed off to the far end of the dyke while Tom walked the length of the dyke towards him.

By now it was 4.30 in the afternoon and in the fading light the pheasant finally fled from the approaching Tom directly into the path of Henry who waited calmly until he could shoot it. They had spent the whole afternoon chasing a single cock bird and finally bagged it within 100 yards of where they had first spotted it. No doubt it made all the more tasty a meal because of it!

They had some wonderful stories to tell and I remember Tom explaining to me how he would feed partridges along a furrow in a field. That way he could down the whole covey with a single shot!

The third brother, Bob, had something of a wanderlust and

worked his passage several times to New Zealand where he would spend time working on sheep farms, saving enough for his passage back to England. Then he would start all over again.

He was to make several of these trips and no one in the village was ever quite sure when he would turn up again. But turn up he did and every time he quickly returned to pot-hunting.

The first time I ever met him he was with his nephew and the pair of them were out in the lanes of Lincolnshire. He had a loaded 12-bore shotgun strapped to the cross-bar of his bicycle just in case he came across anything worth shooting on his travels. Visitors to his home later told me they saw the same loaded 12-bore placed on the mantelpiece above an open fire. The idea, he explained, was to keep the cartridges dry.

In my many years as a gamekeeper I was often asked by farmers if I had ever seen Tom Parker shoot. When I said that I hadn't the reply was usually the same: "By gum, he doesn't miss many." My own feeling was that they had often seen him going home with game but hadn't exactly seen how he came by it.

However, as Tom got older (he would have been in his early 50s) he asked if he could come beating on my shoots. I gratefully accepted for here was a man who knew the countryside and we were to become great friends.

By attending large shoots he picked up a few new tricks and, as a result, started feeding a dyke at the end of his smallholding. He then invited a couple of the lads on the shoot to go with him as he had a few mallard flighting into the dyke where he had been leaving the food. As they lined up along the dyke five mallard circled round and landed in the water close to where Tom was. But not a shot was fired and the two lads couldn't understand why Tom hadn't taken his gun to the ducks as they landed. Eventually two shots did ring out and some of the ducks were seen flying away and Tom asked for the help of a dog in retrieving the two birds he had brought down. When he was asked about the delay in shooting he explained he could see the mallards in the water but he was waiting for all five to swim together so he could account for them all with a single shot!

As time went on Tom attended hare and cock pheasant shoots and, although he was something of a legend in the Marsh villages where all his shooting was done at sitting targets or those going away from him, when it came to driven shooting he was only an average shot.

I have two great lasting memories of Tom. At North Ormsby we sometimes had a white hare or two on the estate. At one hare shoot I warned everyone not to shoot a white hare should they see one. Unfortunately, Tom didn't hear me and at the end of the first drive came back carrying our white hare.

Then, when he must have been in his 70th year, I was standing behind him on a cock shoot in Welton Vale. He missed a couple of good cock pheasants when along came a virtually stratospheric hen bird. I identified it early and, as it was probably the last bird in the drive, I shouted to Tom he could have it. He put up his gun and fired a single barrel and I was amazed when the hen dropped stone dead at my feet. It was possibly the highest flying pheasant I have ever seen shot.

LAST OF THE PLOVER NETTERS

The Parker brothers were also involved in the netting of plovers on the fields stretching from Alvingham to the coast.

This took place on flooded grass meadows and well grazed turnip fields where the sheep had paddled the land and it had subsequently flooded. These type of fields were much favoured by both green and golden plover, both of which commanded high prices, particularly during the Second World War when large numbers were sent by train from Lincolnshire to the London markets.

During the winter months some Marsh farmers netted plover professionally, digging islands in flooded grass fields on which to set their clap net which was then attached to a long line which led back to a hide in a nearby dyke. The idea was to wait until enough birds congregated under the net and then pull the line. This released a trip switch which allowed the net to fall over the birds, trapping them.

Some of these netting sites were permanent and Tom would help the different owners of the various sites in the area. It was a day-long operation and the hides were sometimes manned for several days at a time, depending on the migration patterns of the plover.

A good day's plover catching could result in a bag of 300 birds and with each plover fetching several shillings in wartime, a well run netting site would produce a tidy income for the owner.

Although the netting of plover was legal, the use of tethered live decoys was not. This led to a lot of secrecy amongst the plover netters because they didn't want the authorities to know too much about their activities.

All was fine during the war years as the netters always referred to the plover as 'wipes' or 'pie-wipes' (an abbreviation of 'pied', e.g. black and white) in conversation amongst themselves. This is the origination of the well-known Lincolnshire name 'Pyewipe', a name shared by numerous farms across the county, obviously areas much favoured by plover in the past.

By using illegal live decoys, the netters were able to obtain good bags but by the end of the war several wildlife groups were seeking to have plover netting stopped and tried unsuccessfully to catch some of the netters using live birds. One young lad involved in helping his father run a netting operation was interviewed by the police and completely baffled them by saying all he was after was 'wipes', certainly not plover.

Bird protection groups tried to find the sites being used for netting but the farmers were very secretive and, although the police had to act on information they received, they did not seem too keen to pursue the matter.

By being careful the netters carried on until 1948 when they had some bad luck. One of the older netters was not having much success in his fields so he decided to join the young lad who the police had spoken to on the site he was helping run. While he was helping out the old man suddenly collapsed and died and the police had to be called to supervise the removal of the body. This alerted the authorities to what was going on and the police were obliged to look into the plover netting operations going on in this part of East Lincolnshire. And while they were removing the old man's body from the field they found some live decoys, although no one would admit to ownership of them.

All this meant the police had irrefutable proof that live birds were being used to lure the plover to the trap sites and the information they gathered helped the Royal Society for the

Protection of Birds push through a ban on plover netting, thus ending what had been a lucrative trade for the men of the Lincolnshire Marsh for centuries.

SILHOUETTE ON THE MARSH

Another man who lived on the Lincolnshire Marsh was Johnny Ireland, whose home was in Marshchapel. Johnny was a poacher right from the time he was a young lad and always had his eyes open for the opportunity to make a few shillings.

Once when he was out duck shooting nothing much was happening when over flew a heron. Up went Johnny's gun and down came the heron. He later related the story in one of the local pubs he used and told his eager audience that once he fired, the heron came down like a pair of trousers!

One evening when Johnny was out regaling the locals with his yarns, his son Edwin and another 13-year-old, Jim, decided to take Johnny's poaching gun, a single-barrelled .410, and walk the tall hedgerows in search of pheasants.

They walked some miles before finally coming to a tall hawthorn hedge and they shot about 15 birds, as many as they could carry, before setting out for Marshchapel.

As they crossed an open grass field in the moonlight they saw an object silhouetted against the sky line on top of a small bank. Edwin cried: "Get down, Jim, I think it's the gamekeeper!"

They laid flat on their stomachs in the moonlit field with their bag of pheasants, incriminating evidence of their evening's work, at their side, all the time watching the shadowy figure on the bank and trying to make themselves as small as possible.

They lay like that for a long time before, finally, the object of their attention finally moved. It was only then that they realised they had not been looking at one of the local game-keepers, but the rear view of a grazing pony!

After that scare, they were extra vigilant on the way home and when they arrived back in the village decided to hide their bag of pheasants until the following morning when they sold them to the local butcher for half a crown — twelve and a half pence each in today's money.

Cartridges were hard to come by in those days and a few nights later, as Johnny prepared to go out on one of his poaching forays, he discovered a box of ammunition missing. He asked Edwin if he knew what had happened to them and his son apologised and said he had used them to go shooting rats.

AN AMICABLE ARRANGEMENT

From the First World War right up until the 1960s, the Lincolnshire Fens were heavily populated with wild pheasant and other game.

Towards the end of this period pheasants were fetching £3 a brace and a poacher living at Little London, near Spalding, named Ted Peck, virtually made a living out of what he could poach. His other specialities were eel netting, plover netting and mole catching.

He was a wily old character and his favourite method of catching pheasants was by using a net, mainly on stubble and sugar beet. He went out most nights and only needed a few brace to make his money.

With his expert local knowledge he always raided local farms just before a shoot was to take place. This usually ensured a poor return for the shooters for, if Ted hadn't caught all the birds, he'd certainly frightened a lot off.

One of his main targets was the Cooke farm at Deeping High Bank and every shoot day, thanks to Ted, there was a dearth of pheasants. Ted was suspected of being the culprit and one day, while he was cycling along a nearby road, the farmer, Jim Cooke, stopped him and they had what turned out to be an amicable discussion about resolving the problem.

What happened was that Ted was given permission to shoot game on one of Jim's outlying fields providing he kept off the others. This amicable arrangement solved the poaching problem on the main farm and kept everyone happy, a perfect

example of how countrymen could sort things out to everyone's benefit.

KENZIE THORPE

The poachers we have looked at so far were mainly from the Lincolnshire Marsh, an area of the county which appeared to have its own unique culture concerning poaching and the law.

But perhaps one of the greatest of them all was Mackenzie Thorpe, who came from Sutton Bridge, deep in the Fens, and later became well known as a goose guide and artist.

Kenzie, as he was known to everyone, was highly active as a poacher between 1931 and 1956 and during that time was convicted on a total of 29 charges connected with poaching and causing grievous bodily harm. Those charges cost him over £120 in fines (a considerable sum in those day) and four guns, all confiscated by the local magistrates.

All his stories were well recorded in his biography and do not need repeating here, but in any book on poaching in Lincolnshire Kenzie certainly deserves more than a passing mention. But it must be said that most of his activities were well beyond the law and careful scrutiny of them takes away much of the romance surrounding the popular notion of the Lincolnshire Poacher.

He was virtually a full-time poacher from 1928 until well into the 1940s when he began to turn his attention to his other interests, that of being a goose guide and a spare-time artist. By this time his reputation had spread and he was in demand as a raconteur, speaking about his life and times as a poacher and a goose guide.

There is little wonder that Kenzie was caught so many times

for one of his main poaching guns was a 12-bore — a noisier weapon is hard to imagine. A 12-bore can be heard for miles around and, looking back, it belies belief that a man engaged in an activity which relied so heavily on stealth and silence should consider using one as his main poaching weapon.

Kenzie's poaching was mainly confined to a wide area around Sutton Bridge but once he had been caught a few times, he probably got the blame for the deeds of most of the other poachers in this corner of South Lincolnshire.

As he got older and better known, he was caught more and more frequently. Perhaps he couldn't run as fast as he had been able to in his youth or, there again, perhaps he just became more lawless.

Although Kenzie poached to live, in another time when the need to provide food was not so acute he would, in all probability, have still been a poacher because it was the thrill of the chase that he really enjoyed.

It was not just the traditional game birds, hares or rabbits that were his quarry. In 1941 his own records noted that his bag included 22 different species, including a sparrow hawk which he probably sold for good money to a taxidermist. Everything Kenzie shot had a value and he was one of the last great Lincolnshire characters to live by the gun.

DODGING SHADOWS

I was the son of a small-time poacher and countryman. Dad only poached to feed our large family and rarely received any money for the game he took.

As a young lad I often accompanied him on his trips around the countryside from our home in Welton-le-Wold. One of the earliest poaching escapades I remember was against the humble rabbit. Although our part of Lincolnshire abounded with them and farmers were at their wits end over the damage they were doing to crops, it was still a poaching offence to kill and take rabbits without the permission of whoever owned the shooting rights over a particular stretch of land.

Most of the farmers were tenants and, as such, did not have the authority to give permission for others to kill rabbits. This inevitably led to friction between the tenant farmers, who wanted the rabbits killed, and the gamekeepers who were employed to look after the interest of their employer, who was invariably the local landlord.

On this occasion Dad and his friend Eddie had decided to buy a 100-yard long net to tackle some of the rabbits at Welton-le-Wold. The area surrounding the village was alive with rabbits and Dad and Eddie had heard that a well set long net could account for a great many rabbits in a night's poaching.

On the night in question they went to their preferred field, which was grassed and was situated between two belts of woodland. It was a tricky job setting the net and they had got about half of it set when there was a sudden disturbance in one of the bordering strips of woodland. Some pigeons had flown up and Dad and Eddie immediately suspected that the keeper was about, so they hurriedly gathered up their net and the set pegs and made their way home.

Living on the far side of this particular area of woodland were a pair of travelling woodcutters who felled trees for the landlord and then moved on to other areas. Dad had met them on the farm where he worked one day and they invited him and Eddie to visit their cottage for a game of cards. While they were enjoying the game and a bottle or two of beer, the four of them got to talking about the countryside and its ways. Eventually, the conversation turned to pheasants.

Eddie said: "I bet you two lads get the chance of a few pheasants working on your own in the woods." John, the older of the two men, nodded in agreement and said they had seen several in the wood nearby. He explained that they had got hold of a .410 and had planned to do a little poaching, even though they knew the penalty for being caught could be a prison sentence.

He went on to explain that on the night they chose for their illicit work they had gone alongside the wood where they had seen the pheasants and were just about to start searching the bushes for their quarry when they saw a shadowy figure out in an adjoining grass field.

They thought it was the gamekeeper and dived into the bushes, disturbing some pigeons as they did so, before running through the wood and back to their cottage. That experience was enough to persuade them that poaching was not such a good idea, particularly so because, if they had been caught, it would have spelled the end of their livelihood as well as a possible stretch in prison.

Eddie looked at Dad and the pair of them started to laugh. When they finally stopped he explained to John and his mate of their plans to long-net that field on that same night and, after comparing experiences, it appeared that the two sets of poachers scared each other off while the local gamekeeper was probably tucked up in bed!

KING OF BEASTS

Towards the end of the war when I was about five years old, I used to go out with my Dad and Eddie in our car. Now in those days it was rather unusual for a farm worker to own a car but for as long as I can remember we always had one. The first one I can recall was an Austin Seven, then came an Austin Ten followed by a Hillman.

We were out in the Hillman one Sunday on our way to visit a farmer at Ludford. Dad and Eddie had put a couple of 12-bore shotguns in the car in case we saw anything on the way. Sure enough, we were going along a lane about a mile from the farm when we spotted three pheasants crossing the road in front of us and walking into a field.

Dad pulled the car up and Eddie immediately loaded the 12-bore and shot two of the birds before running into the field, picking up the two birds and jumping back into the car. He then pulled up the back seat and dropped the two birds into the cavity under the seat cushion before carefully replacing it. The birds and guns safely stowed, we continued on our way the short distance to the farm at Ludford.

When we arrived, Bill, the farmer we had gone to visit, came out to greet us and asked Dad if he had any guns with him, knowing full well he usually carried at least one in the car. Dad told him that both he and Eddie had brought their guns whereupon the farmer asked if they had heard a couple of shots down the lane.

Quick as a flash, Dad said they hadn't but they had spotted

a vehicle going in the other direction. At this point I piped up in all innocence and started saying something about pheasants. Dad stopped me and told me to get back in the car, turning to the farmer to explain I had been out with them, the day before and was mixing up what I'd seen then with what had happened today.

Bill the farmer was no fool and knew full well that both Dad and Eddie did a bit of poaching on the side and was happy as long as it wasn't on his farm. He said he was going to see his brother, who managed a farm near Caistor, and invited us to go along with him.

Now from what I can gather, there was a thriving black market in wartime Lincolnshire and when we arrived at the farm near Caistor we were greeted by the farmer who pointed out to us a large herd of deer grazing on winter corn alongside a nearby wood. Joe, the farmer, said he was fed up with the deer damaging his crops and he only wished his landlord, the Earl of Yarborough, would do something about it.

Bill told his brother that Dad and Eddie had their guns with them and explained how they would love to help him out by shooting some of the deer. They asked if there was any likelihood any of the Earl's gamekeepers would be about and Joe said he didn't think there would be.

He was obviously keen on the plan and went to fetch his brother a gun and some ex-Army ball shot and asked his foreman if he would lend a hand too. The plan was that Joe and his foreman should walk to one end of the wood and drive the deer towards Dad, Eddie and Bill.

Dad drove his car towards the wood and the herd of deer immediately ran into the cover of the trees. Dad parked in one of the ride-ways, leaving me inside, and the three men and their guns went off down the ride to lay in wait for the deer.

After some time the peace of the wood was broken by the sound of branches breaking and a number of deer came crashing through the wood and headed straight for the car. They veered round the vehicle and were then followed by a lot

43

more, the whole herd passing within twenty yards of the car.

I was terrified by all this. To a five-year-old a deer looks like a very big animal indeed and I have to admit I burst into tears. In fact, I was wailing at full pelt when Bill, Eddie and Dad returned to the car. They asked me what was wrong and when I told them they cursed their bad luck and decided to make a hasty retreat back to the safety of the farm yard.

It appeared that the herd of deer had got wind of Bill and had turned tail and run for it, coming straight past the car, scaring me half to death in the process.

To my knowledge, that was the only time Dad and Eddie attempted to turn their hand to deer poaching, although you could see the attraction in it for them. Venison was highly prized in those days in a meat-starved Lincolnshire.

After saying farewell to Joe, we headed back to Ludford, Bill sitting with me in the back seat, blissfully unaware of how near he was to two of his missing pheasants!

A FEW FOR THE POT

At the end of the war Dad acquired some legal rabbit-catching rights from his employer. He was allowed to catch rabbits in the fields and hedges while the landlord retained the rights to the rabbits in the woodlands for his gamekeeper to catch. Dad now had the unusual luxury of having the law on his side when he went in search of rabbits.

One morning as he and Eddie were starting to use ferrets I arrived to provide what help I could. By now I was about seven and carried a hazel nut stick, which was quite handy for killing a rabbit as it was being taken from a purse net.

As I approached Eddie called out: "Here's just the man we need. Lend me your stick, lad." I duly handed the stick over and then Eddie asked me to follow him and we walked over to where he had been setting some nets earlier. He then raised the stick above his head and carefully walked round a small clump of rushes. I had seen him do this before and thought he must have seen a sitting rabbit.

Suddenly he struck into the rushes. There was a lot of flapping of wings and then out rolled a hen pheasant, flapping and jumping all over. I was amazed. Eddie handed the stick back to me before grabbing the bird and popping it into his sack. He then made sure he picked up as many feathers as he could so that if they had a visit from the keeper he would have no idea how one of his birds had met its end.

One Sunday morning around the same time I remember Dad coming home with a hare. It was quickly skinned and put

into a bowl of water. As it had only been dead a few minutes, once in the water I could see its muscles still moving.

Dad seemed a bit worried and kept looking anxiously down the road. What must have happened is that he had been out for a spot of early-morning poaching, had shot the hare quite close to our cottage and suspected he had been seen either by the gamekeeper or the village bobby.

He quickly buried the hare's skin in the garden and put the meat in the oven and, by lunch-time, we had eaten the evidence of his early morning activities. So much for having well-hung game!

Eddie's favourite weapon for silent poaching was a catapult. He was an excellent shot with it, using home-made lead balls or ball bearings. He could hit an old penny on a post at ten yards with consistency and, as sitting rabbits or roosting pheasants were usually nearer than that, Eddie and his catapult were a formidable poaching combination.

During his work on local farms he usually carried the catapult with him and, although he never accounted for large numbers with it, if the opportunity presented itself for the odd bird or rabbit, then the usual outcome was another one for the pot, acquired, of course, in the quietest way possible.

His favourite ruse was to keep a careful watch on the bushes used locally by pheasants for roosting and then to pay them a visit on his way home from work when he was usually able to bag one or two.

Eddie lived next door to my grandma so if he had had some particularly good luck, some of his spoils landed on her table. Being a countrywoman through and through, she loved all types of game and, besides Eddie's offerings, granddad brought home partridges which he had picked up on his way to and from work.

In those days the Lincolnshire Wolds were full of them and the ones he collected were those which had flown into the overhead telephone lines alongside the road or those which had been hit by passing vehicles. Whether this actually counted as poaching, I'm not sure, but being in possession of game without a licence or authority must have been regarded as being outside the law.

As well as collecting his partridges, granddad was also a very good hand with a stick and, as was his habit, he would go for a stroll down the lane most Sunday mornings. No one bothered much about what he was doing there: it was just old Bill out for his Sunday morning walk. But once he was out of sight, he began to look for sitting rabbits along the wide grass verges. He was very handy with that stick and there were not many Sunday mornings when he failed to come back without at least one rabbit tied to his braces and concealed neatly by hanging down inside his trouser leg.

When I was young I used to ask if I could go with him but he always said he wanted to be on his own. It was only later I found out why.

TESTING TIME FOR THE MECHANIC

Mention of the roads and grass verges around Lincolnshire brings to mind the story of a local garage mechanic who travelled the roads on the Wolds testing cars.

Often he went at night when the roads were full of rabbits, hopping from side to side in the beam of his headlights. He was a member of a large family and meat was hard to come by in the 1940s so, naturally, he tried to run down as many of the rabbits as possible in an attempt to help keep the family table stocked.

In the beginning he wasn't too successful. Everyone knows if you run over a small rabbit with the large wheel of a car you're going to end up squashing the rabbit and what's left generally isn't edible, at least by humans.

One thing he did notice was how many rabbits escaped by getting under the car and between the wheels. This gave him an idea. He reckoned if he suspended a length of chain across the car between the two front wheels any rabbits which went under the car would be hit by the chain and killed and, like as not, could be picked up for the pot.

He tried it on his next outing and the chain worked like a dream, every rabbit that went under his car was killed. All he had to do was to get out and pick them up and drive on. There was no noise, no going on land, in fact nothing to alert the local gamekeepers.

A quick ride round the Wolds could often net a bag of ten rabbits and they were easy pickings. This went on for some

time and by now the mechanic was getting more rabbits than his family could eat so he began selling them and, not surprisingly, people began to talk.

It wasn't long before the local policeman got wind of what was happening and, out on patrol one night, he saw a car approaching in the distance and it appeared to be stopping quite frequently. As it finally came into view he noticed sparks coming up from underneath the vehicle and so he flashed his torch for the driver to stop.

When he did so he told the driver he had seen sparks coming from under the car and then, when he examined the vehicle, he found it was caused by a length of chain strung between the two wheels. It was sagging and coming into contact with the road, throwing up a shower of sparks as it did so.

By now highly suspicious, the police officer asked the driver why he had appeared to have stopped so frequently. The driver explained he was testing the vehicle and something appeared to be wrong with it and he had kept stopped and getting out to see if he could trace the fault.

When the police officer mentioned the chain, the driver had an explanation for that, too. It had been used to tow the car in for repair and he had forgotten it was still attached to the front axle.

The policeman was still suspicious and shone his torch inside the car but he couldn't see anything untoward and finally he told the driver to be on his way. What he had done, however, is to overlook a tool box pushed underneath the front seat in which four freshly killed rabbits were concealed.

It was a narrow escape for the mechanic, so narrow that he decided that his ingenious method of killing rabbits was asking for trouble. So, when he got back to the garage later that night, he removed the chain and never used it again.

MY POACHING DEBUT

By the age of 13 I had the use of my uncle's .410 and it was my intention to make good use of it.

In the winter of 1953 we had quite a heavy snowfall and looking across Dad's vegetable garden I saw the fresh tracks of hares which came to feed on the cabbages after dark. I asked my Dad if I could wait in the old pig sty nearby and see if I could shoot our nocturnal visitors with the gun. Dad thought it was a good idea and that night I settled down in the sty to wait.

It was a bitterly cold night with a sharp frost and I'd been there, shivering in the sty for some time before a hare finally appeared, bounding across the snow. Although it was dark I

could see it silhouetted against the snow before it finally approached the cabbages. When it was about ten yards away I fired and the .410 killed the hare instantly.

I carefully made my way from the sty and picked up the hare and took it back to the sty and settled down to wait yet again. After some time a second hare appeared and was soon greeted with another blast from the .410. In the space of a short time I had bagged a couple of hares, almost as much as a 13-year-old could carry.

When I arrived home I was full of pride as this was my first solo poaching expedition and, as the object of the exercise was to put meat on the family table, I had certainly exceeded my expectations. Mum was pleased but said we would only be able to eat one of them.

The following day the roads were tar too icy for me to travel into Louth by bus, where I was a pupil at the grammar school. So I made myself useful by wrapping the second hare in a sack and tying it to a small sledge and I walked three miles across the fields to my grandmas house. She was very pleased as hare was her favourite and I was delighted for my first poaching expedition had managed to feed two families.

The following winter I was out after pigeons with the .410. One day I was in a thick fir wood where there were also a few larch trees. As dusk approached a few pigeons were flying

around, looking for somewhere to roost, but they were not being very co-operative and always seemed to be just out of shot.

As it got dark I heard a flock land not far away from where I stood and, not having had a shot, I thought I would try and stalk them. After moving a few yards in their direction, some of the birds flew away but I could still see some dark shapes silhouetted against the gathering gloom in one of the larch trees.

Settling myself, I took aim and fired and a body fell near me. At the sound of the gun more birds flew away but at least two remained in the larch, so I reloaded and fired again and another bird fell close by. When I picked it up I realised it wasn't a pigeon I had shot, but a hen pheasant. When I found the first bird, I discovered it, too, was a hen pheasant.

When I looked up the third bird was still sitting there so I shot that as well and went home with three pheasants. When I arrived the greeting my news received was not too enthusiastic. For although the pheasants made good meals, both Mum and Dad could see my game killing could lead to trouble, particularly if I got carried away and went after more.

After that night I was told to leave game alone and restrict my shooting to rabbits and pigeons. Although in all honesty, that's what I thought I'd been shooting that night... at least, the first two.

SNARING A BONUS

When Dad first obtained permission to catch rabbits on the fields, trapped or snared rabbits fetched eight shillings a couple. This price was usually fairly stable throughout the winter but, on occasions, could rise to nine shillings. To put this into perspective, at the time Dad was working a 44-hour week on the farm for just £5.

One Saturday afternoon he set his six dozen snares and that evening and the following morning caught 11 couples of rabbits. That brought in £4 19s so, for a night's work snaring, he made virtually a week's wage. At the time hares were fetching five shillings and, while snaring, he sometimes illegally caught a hare and, so long as he wasn't caught, he made a little extra on the side.

By this time Dad had a Ford 10 van and, because there was a rail strike on, a local fish merchant asked him if he would take a day off from the farm and deliver a consignment of fish for him to Lancashire.

This involved a very early start and meant he could not inspect all his snares before he went. As they had been set a few nights he said there shouldn't be too many rabbits in them and, as it was the school holidays, I could go round them at first light for him.

I had about two miles to cycle to check the snares, which were spread across two fields. I got three rabbits in the first field and when I went into the next one, the first thing I found was a dead hare in the first snare I came across. I quickly took it from the snare and hid the body in a hedge bottom before going round the rest of the snares and collecting two more rabbits.

I had no bag or sack with me as the recognised way of carrying rabbits on a bicycle was across the handlebars or crossbar. As I collected the five rabbits I wondered how I could get the hare past the keeper's house which was on my route home. I decided to take the rabbits and leave the hare where it was for the time being and when Dad came home I told him what I'd done.

He said I'd done exactly the right thing and, although he'd just driven to Liverpool and back, he jumped straight into the van and took me to show him where I'd left the hare. By this time it was dark and starting to snow but we managed to find it and we went home for tea, the hare well hidden in the back of the van.

A TALL STORY FROM
THE LIKELY LADS

Years later my head-keeper caught some young lads ferreting rabbits and asked them to leave the estate, even though they claimed all they were doing was taking their ferrets for a walk. He told me the reason he took this action was that experience taught him that young poachers soon graduate from fur to feather.

I couldn't argue with this for that's exactly what I had done years before. And just to prove his point, the lads in question, four brothers, soon moved on from just keeping ferrets to having lurchers and then guns.

They lived in the local town and had found work on a poultry farm. They soon asked permission to shoot rats on the farm and the farmer readily agreed to anything which would help rid him of this nuisance.

This gave the lads a reason to be there after dark and it just so happened it was only a short walk from their poultry farm to one of my pheasant woods. Although I never caught them, I knew they visited the woods because one of them was a very small lad and I sometimes found a very small footprint on the feed ride.

They were caught several times by different farmers and keepers, usually for ferreting or hare coursing, but their undoing came in a rather strange way.

They were not satisfied with making do with the permission

they had had from their employer to shoot rats and one night, quite late on, they decided to 'borrow' his car and travel the local roads to do some rabbit shooting with a .410.

I'm not sure precisely what happened, but about two miles from the farm the car crashed through a hedge and was badly damaged. They then walked back to the farm at about midnight, knocked the farmer up and claimed that while they had been out lawfully disposing of his rats, someone had stolen his car. The farmer rang the police and during the call they asked the lads which way the car had gone. They duly gave directions and, in the meantime, offered to accompany the farmer to see if they could locate the missing vehicle.

After a short drive the farmer found his car embedded in a fence and it was at this point that the police arrived. The police officer concerned was well aware of the lads' reputations as local poachers and, taking the farmer to one side, said he suspected they had taken the car.

When he began questioning them they changed their story and were later charged with taking the car without the owner's permission.

Although their subsequent conviction didn't curtail their activities, it showed them up in a different light to the local landowners who had given them permission to catch rabbits on their land. They had used this as an excuse to go after game and, if they were ever stopped, all they had to do was to produce the rabbits they had caught on land they were entitled to be on.

OPPORTUNISTS
ON A MOTOR-BIKE

When I was young I went beating for two local shoots and one day when we were out on Mill Farm at Welton-le-Wold I witnessed a very unusual way of taking game without permission.

The gamekeeper and other beaters were working a turnip field towards the guns, who were standing in a small valley. As the pheasants and partridges went over the guns some of those shot dropped over the brow of the hill.

While this was going on I spotted a motor-cyclist coming down the road which turned at right angles shortly after the point where the guns were standing.

There were two men on the motor-bike and, as they approached the bend, the pillion rider jumped from the machine and ran across the corner of the field, picking up as much game as he could carry. In the meantime, the motorcycle had continued along the road and went round the corner before stopping to let the pillion rider jump on again before the two men made good their escape, complete with several pheasants and partridges.

We watched all this unfold from the other side of the valley. Had we not been on the high ground at the time the whole operation would have gone undetected.

Needless to say, the two men were never identified or caught and it all went to prove that sometimes all you needed was a spot of quick thinking and strong nerves to get away with some illicit game-taking.

HAROLD'S EXPENSIVE RABBIT

At the age of 15 I left school and got a job as a gamekeeper for the Nickerson Brothers at North Ormsby so my transformation from poacher to gamekeeper was quicker than most.

Nickersons also owned the world-famous seed company and employed a number of lorry drivers, one of whom was called Harold.

One night Harold's lorry ran off the road and shed its load of corn into a ditch. Harold was uninjured and left his vehicle, hitching a ride into nearby Grimsby with a passing motorist.

The next morning another lorry driver named Ben was told to accompany Harold to the crash site and transfer the spilled sacks to his vehicle before helping get Harold's lorry back onto the road.

As they lifted the last sack onto Ben's lorry, they discovered the remains of a rather squashed rabbit on the roadside. Harold spotted the rabbit and paused before saying: "I just knew I'd got that rabbit!"

CAUGHT RED HANDED

In my young days as a gamekeeper we had a lot of trouble with roadside poachers who drove alongside fields shooting at game from their vehicles.

It wasn't long before I caught a couple of lads on a farm road at North Ormsby. They had a lorry and were doing contract work on the farm and, as I drove up unnoticed, they had just shot a cock pheasant with an air rifle.

I picked up the pheasant and asked for their names and addresses. They gave them to me and both had home addresses in Central London. I told them I would be reporting the matter to the village policeman and later he went to where they were both working — they were painting a barn at the time — and took statements from them.

My head-keeper, Ernie Turner, said we should make a case out of it and prosecute them as, although they were clearly naïve and not really the calibre of poachers we were looking for, it would serve as a warning to the real culprits. This we did and both were later fined £5 in their absence at Louth Magistrates Court when they pleaded guilty by letter.

Within a month we were lucky enough to catch two more. This time Mr Sam Nickerson, Ernie Turner and myself were riding across a stubble field looking for partridge when a car passed us on the distant road. We noticed the rear window of the vehicle was open and watched it turn onto our shoot at the cross roads.

We let it go out of sight and then followed at a discreet distance. As we rounded a bend the car had stopped and one of the occupants was out in a roadside field picking up a fluttering cock pheasant.

Ernie and I jumped from our Land Rover to question the driver of the car and his accomplice as he returned with the pheasant in his hand.

They held their hands up and said it was a 'fair cop' and we stayed with them on the roadside while Mr Sam went to fetch the local policeman. Remember, these were in the days before mobile phones!

While he was away we got talking to the two poachers and they asked us who the man in the Land Rover was and when we told them who he was one of them told us he was having a particularly bad week as the day before he had lost his job with Mr Sam's brother, Joe, when he had been searched leaving work at Cherry Valley with a car boot-full of Mr Joe's frozen ducks!

Despite their misfortune, the two men bore us no malice and even took us for a drink after they had also been fined £5 each by Louth Magistrates.

A REAL WILD GOOSE CHASE

One Christmas a farmer at Welton-le-Wold had some geese which he planned to kill and sell on the local market.

On the appointed day he began rounding up the geese but the windy conditions were hampering his efforts. As he tried to catch the geese one of them became airborne and, with the help of a particularly strong gust of wind, managed to fly across the farm yard and eventually disappeared from view in the general direction of the village.

Later, after dealing with the remaining geese, the farmer went into Welton-le-Wold to ask whether anyone had seen his missing goose. No one had but each time he went back into the village over the next few days he continued to inquire whether the bird had turned up.

At this time Dad kept a few geese of his own as a way of ensuring we had a good Christmas dinner. One night he went along to a whist-drive in the local village hall. The farmer was also there and during the night he accused my Dad of adding the missing goose to his own small flock.

This was particularly upsetting for Dad for he was the only person in Welton-le-Wold who actually knew what had happened to the goose. It appeared the wind had actually taken the goose some three miles from the farm and, over the following few days, Dad saw it grazing on the farm on which he was working. He had tried to approach it but was very wild and fluttered away whenever he got anywhere near it.

However, after being virtually accused of stealing it he

thought he might as well put the record straight and the next day he took his 12-bore and shot the goose,

Originally, he had intended to tell the farmer where it was but, after being accused, he didn't want anyone else to get the blame. After all, if you're going to get blamed anyway...

PARTY TIME AT STENIGOT

On the Stenigot Estate was a shooting lodge with a cosy fire-place where the shooting parties went for lunch.

It was situated in the middle of the woods and one of the local keeper's duties included looking after the lodge, lighting fires on shoot days and generally taking good care of the place.

One evening after shooting, the guns remained in the lodge talking amongst themselves about the enjoyable day they had had. The keeper went home for his tea and when he went back to the lodge, he found they were still there so he decided to go to the local pub with the intention of going back later to tidy up and check the fire was out.

He spent a convivial night in his local and, when it was time to leave, suggested to a few of his friends that they should go back with him to the lodge and keep the party going.

When they got to the lodge they found it was empty but the fire was still burning and it was pleasantly warm inside. They had brought a few bottles with them and, in no time at all, the party was in full swing.

By now the keeper was decidedly merry and decided it would be a good idea to invite the landlady from the pub to join him for a dance on the table. Everyone was roaring with laughter at the sight particularly when the hapless keeper and his partner fell off. The keeper, however, found it far from funny. He was a big man and injured his ankle in the fall and he had to be carried out of the lodge and put in the back of a car and driven to hospital for an X-ray, his friends fearing he

had broken it.

Some of the others tidied the lodge up and made sure the fire was out before leaving. In the meantime, it was discovered the keeper had suffered a badly sprained ankle and when he finally returned home he would have to be laid up for a few days.

The following day the keeper's employer went to see him after hearing he had had an accident. He particularly wanted to find out how he was and how he had come by his injury.

The keeper managed to keep a perfectly straight face when he explained that, as he was locking up the lodge, he had heard the unmistakable sound of poachers in the wood and had gone after then, only to trip over an unseen rabbit burrow and fallen badly. This seemed to satisfy his boss who, no doubt, went away thinking what a truly conscientious man he employed!

A WET NIGHT'S POACHING

I was the classic example of a poacher-turned-gamekeeper but in the 1950s we had a police constable in Welton-le-Wold called Alan who was a poacher-turned-policeman.

He made a wonderful village bobby because, having been a poacher himself in the past, he knew the ways of the poachers in the area and how they operated.

One day he told me about his life before he finally joined the Lincolnshire Constabulary when he and his pal George poached lots of the farms across the Wolds between Wragby and Louth.

He was a big man even by the standards of the police force in those days and I still recall the amusing tale he once told me about a night they spent poaching on the Hainton Estate.

He and George found some roosting pheasants in some thorn bushes which surrounded a small pond. If pheasants have the opportunity they will always roost over water and this turned out to be Alan's downfall that night.

The two men crept round the bushes and, although some of the birds managed to fly away, they did shoot nine between them, all the birds falling into the pond. One or two were quite close to the edge but the remainder were in the middle.

They collected those within reach and then, with the aid of a long stick, began trying to pull in those in the centre of the pond. It was at this point that Alan over-balanced and fell head first into the water. And it turned out to be a lot deeper than he had bargained for.

The pond was full of stinking, stagnant muddy water and when he finally found his feet he was able to stand up, the water coming up to his shoulders. However, he was still first and foremost a poacher and he made sure he grabbed the remaining birds before making his way to the edge of the pond and hauling himself out.

It turned out to be a long, cold and wet walk home for the bedraggled poacher, even though he had the satisfaction of knowing they had reclaimed all nine of their still-damp pheasants.

A SHOCK FOR THE AIRBORNE
DIVISION

In the 1950s as Lincolnshire recovered from the austerities of the Second World War more shoots were formed and more pheasants were reared to supplement what was a very healthy wild stock. These reared pheasants tended to be tamer than their wild cousins and, subsequently, were much easier to poach.

The Worlaby Shoot south of Louth was on one of these pheasant-rearing estates and quickly began to attract the attention of poachers and the men concerned had a unique way of finding where then birds were roosting.

One of the ringleaders had a private pilot's licence — he was clearly not the average run-of-the-mill poacher — and before a poaching foray would circle a particular area of the Lincolnshire Wolds in his private plane in the late afternoon so that he could observe exactly where most of the pheasants were congregating to roost and would make a note of which woods were heavily stocked in readiness for his night's work.

Later, he and his two brothers, Norman Elvin (more about him later) and his brother and a friend would be taken to the area by his sister in a large Austin car and dropped off after arranging a pick-up point for later in the night.

The group of poachers then set off into the woods, two of them shooting with .22 rifles, two picking up and the other two on look-out and carrying outside the wood.

On this particular night they moved along a wooded valley bottom and soon collected two bags of pheasants which were carried up the hill to a distant road to be picked up later in the night. They then returned and continue shooting.

As they were fairly close to the keeper's house, one of the poachers carrying a rifle bumped into an electric fence erected

to contain stock. I can't think of a better electricity conductor than a rifle barrel and, as sparks flew, the severely shocked poacher dropped his gun, thinking momentarily he had been caught by the gamekeeper.

He quickly recovered his composure and his rifle and carried on, collecting more sacks of pheasants before being picked up. Their car was then used to collect the bags of pheasants which had been left at the roadside. When they arrived back at the brothers' farm the count was 140 pheasants, hardly a case of going out to get 'one for the pot'.

One of the men involved lived in Barnsley and he had a ready market for every bird they could poach.

YOUR JACKET, SIR?

Some time later the Worlaby Estate was subjected to another spate of poaching. After many nights out investigating, the keepers had some idea of the areas favoured by the poachers and one night one of them lay in wait as a pair of poachers headed straight for him carrying a sack.

Just as the men neared him, the keeper jumped out to confront them. Both men turned to flee, one dropping the sack as he did so. The keeper grabbed him by the jacket and, in the ensuing struggle, the man shed the coat and ran off, leaving the keeper hanging onto the jacket.

Later it was discovered the sack contained the bodies of 21 pheasants and when the jacket was examined it was found to be covered in lime dust. As there was a quarry on the edge of the estate, it was fairly obvious where the poachers had worked. Although no convictions could be brought on this evidence, it was fairly clear to everyone who the culprits were.

RISE OF THE LINCOLN GANG

By the early 1970s we were rearing some 7,000 pheasants on our shoot at North Ormsby and we quickly began to attract the attention of the nocturnal poachers.

They were far harder to catch than those who tried their luck during daylight hours and, in one particular season, we were being visited by them roughly once a month.

One of the other keepers, Steve Nicholson, and myself did everything we could to prevent them but, after having no trouble on the estate for such a long time, we had to start from scratch to combat the problem they posed.

The Poachers' song contended: 'It's my delight on a shiny night...' but this just wasn't true. Moonlit nights were too bright for poachers; the birds could see them coming and would fly away on their approach.

In the early stages we found the visits usually occurred when there was a half moon and it was just going down, with two or three o'clock in the morning the favoured time.

After several visits, which we first realised had happened when we found small clumps of feathers left under the thorn bushes the pheasants used for roosting, we managed to establish that the poachers were walking about two miles to this particular wood when the conditions were right. This meant we did not have to be on patrol every night, unlike some other Lincolnshire estates which ran a virtual night-shift for their keepers.

We could never tell exactly how many birds we were losing

but, with the poachers having to walk so far and part of it uphill, we thought it was nearer 40 than some of the figures we were hearing from other estates. With the average hen or cock bird weighing over three and a half pounds, around 40 would have been as many as two men could carry comfortably over a two mile walk.

The following season we received a visit from the poachers in October, a little earlier than usual as there were still a lot of leaves on the trees, something which made spotting roosting pheasants difficult. However, the visit was to follow the same pattern as the previous year and this put us on alert.

Whenever conditions were 'right' for the poachers, Steve and I would take it in turns to spend most of the night on patrol, the plan being that if we saw anything we would first of all contact the police and then the one of us not on duty.

After a few nights of this I was woken by Steve at about 1.30am and he told me he had been using his night lamp to scan the old bomber airfield at Kelstern and had seen what appeared to be a caravanette protruding from the remains of one of the derelict Nissen huts. He had immediately pulled the beam away and made it look as if he was continuing his search. He had then gone home and called the police before coming round to collect me.

We went straight away to the airfield and waited until the police arrived and when they did Steve told them what he had seen. We knew that a particular gang of poachers from Lincoln used a blue and white caravanette and, as we approached the hut, we could see it was a similar vehicle. The police officers told us they would have a look at the vehicle and then stake it out and wait for the occupants to return.

However, we were the ones in for a shock. As the police officers shone their torches through the window of the caravanette we could see the enormous figure of a man inside shielding his eyes against the light. We had always suspected one of the poachers was a big man because of the size of the footprints we had found on previous occasions. However, as he gradually unfolded from the vehicle he was even bigger than we thought! His accomplice turned out to be of more average build, but we were all certain these were the men we had spent so long trying to catch.

But if we thought we had caught them red-handed we were to be disappointed. When the police searched the caravanette all they found were a couple of lunch bags filled with sandwiches, two flasks of tea and a couple of empty baskets. When the men were questioned, the men said they were waiting for it to get light before going out to look for mushrooms. We knew this just wasn't true and they were duly taken with their vehicle to Louth police station for further inquiries.

Steve and I searched all around the old RAF buildings but all we could find was a set of footprints which went across a small stretch of ploughing. When daylight came we searched again but could find no more clues about what the men were really up to at Kelstern.

Later in the day we heard the police had searched the caravanette from top to bottom but, after finding nothing, they had had to let the men go after they stuck to their story about mushroom picking.

We believed these two men had poached parts of Lincolnshire for many years but after that night they seemed to concentrate their activities on other estates and left us well alone. They were caught several times but, after being taken to the local police station, were always allowed to go because of lack of evidence.

On one occasion they even taunted the police by telling them what really happened the night we caught them at Kelstern.

They claimed they were just setting out across a ploughed field for a nearby wood when Steve's light fell on the Nissen hut. They were not sure whether their vehicle had been seen so they decided to hide their rifle and their carrying bags, which had been adapted from old mail bags. Their plan was to carry on as if they had not been spotted and to get back to their vehicle before first light.

These men were, we firmly believed, professional poachers who would be out five and six nights a week either side of a full moon on estates all across Lincolnshire.

I related this particular story in my book 'A Covey of Tales' which was published by Ashridge Press in 1999. Later a review appeared in the 'Lincolnshire Echo' and, as a result, I received a most interesting letter.

It was addressed to: 'Head Game Keeper, Demic Mill, Acthorpe, North Ormsby, Nr Louth, Lincs' and was seemingly written by someone who suffered from dyslexia or was seriously bad at spelling. After much deciphering, what follows was my translation.

"For the keeper,

I read in my good paper about your book. The bit about battling with poachers made me think.

We met one night. I think it was 30th October, 1979 on the airfield at Kelstern. Our van was half in an old RAF shed. You came over with the police, remember going to the rotten zinc sheets looking for something? When you first shone your light on us and then went, we took all our gear and planted it on the roadside near the iron gate. We got locked up in Louth but then let go, we collected the gear the next day.

We came back on 4th November, 1979 and planted the van down the same track around a little bit at the bottom and we covered the van with an old tent. Someone came down the track and shone down the road but never saw us. We walked past the house down the hill to your wood. We never bothered with the big wood too much and that night we took 45 pheasants.

We moved our parking down to Utterby village, leaving whatever we had in the middle in a spare one (author: vehicle?) and hung low till later, but still left the van in the cutting only to walk up the track to the wood, which was better than carrying uphill back. We went in most woods, but they were the best! We worked from the factory at North Thoresby to Louth and then moved over to the Alford area.

I spent two nights in Louth within a week, but walked away. We always collected the gun and birds. But when we were over in Alford we came unstuck and got a £150 fine. The police got 32 pheasants but we got the gun and a bag of pheasants back the next day.

That was the last of our bad luck. The snow came a few weeks later and that finished the season, but not before we took 88 from Well Vale to pay the fine with!

I started shooting in 1947 and finished in 1994 because of leg trouble. I have written a book with all the names and places and the bag we took but may never get it published.

That all passed when our team was four men, but we only ever worked with three of them at the same time. Two are now dead and us last two are good for nothing, yet we have our memories of our nights out together. You'll never have trouble with us again. In all our years together I had three convictions for night poaching, one of the others had a few and the younger one had one conviction.

Three convictions in the years 1947-1994 says something. The fines were paid out of the kitty three ways. That's all gone and I hope you don't blacken our names too much.

I wish I could say I'm sorry for the trouble we gave, but if I did I would be telling a lie.

I wish you every good luck with your book, good health and many more years good shooting.

Good luck to you and yours. Sorry, no name. Just thoughts.'

I think this story sums up neatly the activities of what was known as the Lincoln Gang by most of the keepers in Lincolnshire. I believe that just about every well-stocked shoot in the county received a visit from them at some time or other.

The first time I heard about them was when they were caught at Haverholme, near Sleaford. Haverholme was a very heavily-stocked shoot with thousands of pheasants. In the 1950s it was run by the head-keeper, Monty Christopher, and four under-keepers and with this number of staff it was quite easy to run a night-watching rota but even though they were very vigilant, the estate was visited on more than one occasions by the poachers. Once or twice they were spotted in one of the woods on the estate but each time they managed to escape.

On one particular night the keepers heard them in this particular wood and men were positioned on three sides, the fourth being bordered by the River Slea. Once again everything went quiet and it appeared they had got away. However, with the help of the police, a vehicle was located and some of the gang were caught with sacks filled with 83 pheasants.

One of the poachers got away but left his cap behind. This was found by the police and they had a good idea whose head it fitted. Later they went to the man's home in Lincoln, knocked on the door and were greeted by his wife. The police officers asked where her husband was and, although it was still very early in the morning, she said he was out. The policemen then showed her the cap and asked whether it belonged to her husband. By all accounts, she looked extremely shocked and asked what had happened to him. All the officers would tell her was there had been an 'incident' and they would wait for him to return home.

This they did and later he turned up, having walked all the

way home from Sleaford, and on his arrival he was promptly arrested.

The other members of the gang were all caught at the scene and it was only then that it was discovered how close they had come to getting away — and how they had escaped on previous nights.

As I related in 'A Covey of Tales', one of them was an exceptionally large man and on that night he had waded across the River Slea, carrying first their haul of pheasants and then his mates, the water coming up to his chest. This probably accounts for him always appearing to have a cold when he visited our shoot in the 1970s. How did we know this? He left cough sweet wrappers in the woods! He must have been suffering then from some earlier soakings in the Slea.

For that night's work at Haverholme, the big man received three months in prison. It turned out the weapon he was using was a .410 which was fitted with an ingenious home-made silencer made from a baked bean tin and welded to the underside of the barrel. A torch was strapped to the top of the gun making it a very effective weapon for poaching,

THE LONG MARCH HOME

Although the poacher's letter makes it clear the Lincoln gang consisted of four men, others from the city probably heard what these men were achieving and set out to see if they could emulate them.

This must have been the case for there were occasions when sometimes two or three estates in the county would be visited on the same night by different gangs of poachers, keepers often finding cars belonging to suspected poachers left in pub car parks or village streets as decoys. While the police and keepers kept a watch on these vehicles, the gangs, most of which had at least two cars or vans, were busy in other woods,

One of their number went to great lengths to avoid detection, having had an accident with his .410 whilst out poaching one night. He returned home at 2.00am with a serious injury to his thumb, but refused his wife's plea for him to go for hospital treatment. He nursed his injured hand until first light when he told his wife to go into his garden shed and fire the .410. On her return he asked her to call an ambulance and circulate the story around the neighbours that he had been in his garden-shed cleaning his gun, when it had gone off, and he had sustained an injury to his hand. This then disconnected him from a poaching incident which had happened in the middle of the night, because all his neighbours had heard the shot and would provide alibis if needed.

I had evidence of this flexibility shown by the gangs in the 1980 when pheasant prices were still fairly high and most estates were still being poached, some of them very heavily.

I was out one night with the local police sergeant from the Lincolnshire force — incidentally, another poacher-turned-policeman — when a message came through of a poaching incident on the Rothwell Estate, just off Caistor High Street, the Caistor-Horncastle road which runs along the top of the Wolds and was on the extreme boundary of the sergeant's beat. We decided to go and see if we could help in any way.

When we arrived some of the Rothwell keepers and local policemen were talking to a young man in a car while some of the other keepers were using their torches to light up both sides of the small wood where the car had been parked. They told us they had come across the parked vehicle and, while they had been speaking to the driver, had heard the unmistakable sound of men in the wood. They had quickly surrounded it and used their lights to illuminate possible avenues of escape for whoever was inside.

However, when we combed the wood we found nothing but the driver and his vehicle were taken to Louth police station for further inquiries to be made. It was believed that by impounding the vehicle in this way, the other men would be deprived of transport and would have to walk out of the area.

We later heard that there were, in fact, four other men and they had managed to escape before the beams of light were placed alongside the wood. They laid low for some time close to Caistor High Street, leaving their rifle and ammunition hidden on a grassy bank.

When their driver failed to turn up at the agreed pick-up point they decided to walk back to Lincoln across country, using roads only when absolutely necessary. It was a fair old walk for them, the best part of 20 miles as the crow flies. Three of them were young men but the fourth was somewhat older and was the man who had had the accident with the .410 some 20 years earlier.

By six o'clock the next morning the weary poachers were not too far from Lincoln. By now they were using the A46 and the older man in particular was getting very tired. He was not being helped by the gang having to dive into the nearest ditch every time they saw a vehicle approaching.

Eventually the older poacher had had enough and decided to chance his luck, announcing he was going to try to hitch a lift from the next vehicle which came along. The others were too tired to disagree and watched him as he flagged down the very next vehicle. As luck would have it it was a police patrol car, one of those out looking for the men. As it slowed down he realised his mistake too late.

All four men gave themselves up to the single officer in the car and he took them straight back to Louth police station where they were questioned for two days before finally revealing where they had hidden their rifle and ammunition. They had, it seemed, been disturbed before they had managed to do any shooting and were eventually charged with armed trespass but not killing game.

ALARM OVER THE POACHERS' MARKERS

Besides our regular patrols we used special alarm guns in our woods to give protection against poachers.

These were either free-standing metal ones which were on a metal peg and incorporated a heavy weight which dropped onto a firing pin when activated by a stand of wire stretched across a ride or a footpath, or a more modern one which was small and had a spring-loaded firing pin and could be screwed to a fence or a tree. Like the more traditional alarm gun, this was activated in the same way with usually a strand of copper wire stretched over a route attractive to poachers.

Both these types of alarm guns could be set off by a large bird, such as an owl, or by a roaming animal. I have known them activated by the wind and, even though these false alarms did happen from time to time, they always had to be investigated. In addition, we never quite knew how many times poachers had been warned off by springing one of the guns when the keeper was elsewhere.

The reason I mention these alarms is because we felt they gave us some security and protection on the estate until one night when it happened to be my turn to be on anti-poacher patrol.

We had been pre-warned by someone that a gang of would-be poachers was on its way from Lincoln but their eventual destination was unknown. They had a choice of a wood anywhere in the county as they were highly mobile and did not seem to mind how many miles they clocked up in pursuit of illicit game.

It was a terrible night with high winds and lashing rain. I had arranged with a neighbouring keeper that he would patrol early and I would go out after midnight. Conditions were still bad by the time my spell of duty was to begin, pouring with rain and everything battered by that strong wind.

I patrolled over our estate and our neighbours, spending time at all the spots where I was likely to see or hear poaching activity. By 3.45am the weather was still terrible and I had seen nothing so I decided the poachers must have chosen somewhere else for their activities that night and I went home.

After a few hours rest, I was woken at 9 o'clock to be told that the last wood I had inspected had been badly poached. It was a thick blackthorn wood, well stocked with pheasants which loved to roost in thorn bushes. It was also well-equipped with several newly-fitted alarm guns. They were set on the few tracks between the blackthorn bushes and securely fixed in place. The trip wires fitted to these guns were hardly discernible to the naked eye in daylight. Yet the poachers had found every one of them, ripped them off the bushes and had stolen them as well as the pheasants! All this on one of the worst nights I had ever spent out watching for poachers.

We followed the tracks of the poachers back to where they had left the birds to be picked up by the roadside and found a small cash bag from a bank fixed to a nettle stalk by the roadside. This was the Lincoln gang's trademark and we had found similar bags after previous raids on the estate.

A small plastic coin bag looks insignificant enough but it stands out like a sore thumb in car headlights at night when suspended on a twig or a stalk at the roadside along a country lane.

I believe this marker system was used by this particular gang on the occasions when they used two vehicles, the plan being that if everything went well they travelled back to

Lincoln with the gun and ammunition in one car and the pheasants in the second. It was less incriminating this way if they happened to be stopped.

The next deterrent we used was an expensive electronically controlled beam which some estates in Lincolnshire were already using. But again these were not fool proof because they could be tripped by a wandering animal or a bird. However, they did help.

At the height of all this poaching activity, Mr Joseph Nickerson (later Sir Joe), who owned the big Rothwell Estate along with many other farming interests, was so concerned he even hired a private detective in an attempt to track the movements of these poachers.

This coincided with a stepping up of police activity around Lincoln. Known poachers would be stopped regularly by the police, both at night and during the day. Whether this came as a result of tip-offs from Mr Nickerson's private detective I do not know, but it seemed to have the desired effect.

I don't know if the poachers knew they were being targeted like this but shortly afterwards the price of pheasants slumped and their activities no longer proved as profitable as they had in the past.

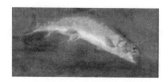

TURNING TO TROUT

For many years the poaching of pheasants had gone on from early October to well into February on those estates with plenty of breeding stock left. After February the poachers turned to trout.

There were several trout farms around Louth and most of these lost fish to poachers at some time or another.

Fortunately, we did not have trout on our estate but the one at South Thoresby was badly hit several times by the same gang who poached their woods in the winter, the same men who were referred to in the poacher's letter. They always left their familiar trade marks — the big footprints and traces of where they sat and ate their pack-up. Poaching can make a man hungry!

To give readers an example of the scale of the problem of game poaching in Lincolnshire I'm reminded of the story of a young fellow from Retford who started game dealing.

He regularly attended the produce auction at Gainsborough market where there were a few pheasants on sale. He made a successful bid for them and was later loading them onto the trailer he towed behind his car.

While he was putting the birds into the trailer he was approached by a man who said to him: "You're new. Are you dealing in game? I saw you bought all the pheasants on the auction."

The young chap explained he had just started in business and was hoping to specialise in game dealing. The man replied: "If it's pheasants you want, lend me your trailer and I'll meet you in two hours with 500 in it — if your price is right. And I can let you have the same amount every week up to Christmas."

Quite clearly the man was a poacher, or closely connected with a poaching gang. Needless to say, the young man from Retford declined the offer but the fact that it was made just went to show how many poached pheasants were around the Lincoln area at the time.

BY THE LIGHT OF … THE STEELWORKS

Until street lights became more widespread along main roads and in villages, my part of Lincolnshire was usually fairly dark whenever the moon wasn't full or when it was cloudy. But with the spread of streetlights into the country there are very few parts of Lincolnshire which get really dark, even in the depths of winter.

The conditions we now experience have been normal for both gamekeeper and poacher in the Scunthorpe area for many years because of the glow in the night sky from the local steelworks.

I used to visit shoots in the Scunthorpe area regularly in the 1950s and 1960s and the keepers were constantly night-watching as the poachers could always see their way around the woods at night, particularly when the molten iron skips were tipped when there would be a particularly vivid glow in the sky.

Poaching was very bad around the villages near Scunthorpe as the poachers could simply walk across the old iron ore workings straight into the woods.

As well as shooting roosting birds, they also used trail nets in the sugar beet fields or on stubble. The problem got so bad that I have seen stubble alongside woodland being bushed with thorn twigs to make it harder for the poachers, a sure sign of trail netting activity.

A trail netting gang usually consisted of three men, two to stand by the poles from which the net was suspended and a third to follow and kill the game trapped in the net. Game caught and killed in the net was used either to help weigh the net down or stuffed into a sack carried by the game killer, depending very much on what type of cover was being netted.

A netting gang was once overheard planning a visit to a sugar beet field at Appleby, near Scunthorpe. The keepers were tipped off and waited at one end of the field. Eventually the poachers arrived and sorted their net out at the far end of the field from the keepers, who could make out the shadowy figures but were too far away to catch them.

They remained in hiding until the poachers made their way along the field towards them. They waited for some 20 minutes until they judged the poachers were near enough and then they broke cover. The poachers quickly saw them and dropped their net before running for it, eventually making good their escape.

When the keepers returned to the net they found the bodies of nine pheasants were already tied to the back of the net as additional weights. This shows how efficient netting can be: in the space of 20 minutes three men had trapped and killed nine pheasants silently, an indication of what damage a netting gang could do to an estate if they remained undisturbed.

The net recovered by the keepers on this occasion was some twelve yards square with a pole at either end and tied top and bottom and was made in a four-inch diamond mesh of light nylon. The method of catching was such that as soon as the poachers felt a bird strike the net they dropped it, one man moving to the front or the net and the other to the back to prevent the bird escaping while the third moved in to kill the pheasant.

DEALING WITH DESPERATE MEN

Although the subject of this book is very much the poacher at work in my own county, the incident I relate here happened in Nottinghamshire but did have strong Lincolnshire connections.

It involved a former head-keeper at Rothwell, Bob Bradley. Bob had been a keeper in Lincolnshire for most of his life and had dealt with poachers in that time who, by and large, relied on their ability to avoid detection but, if caught, would usually say 'fair cop' and go quietly.

However, earlier in his career after being a beat-keeper at Rothwell, Bob moved to the large Welbeck Estate in Nottinghamshire as head-keeper. As Welbeck was surrounded by coal mines and pit villages it attracted more than its fair share of poachers from the mining communities. And Bob quickly discovered that these men were much different and far more persistent than those he had encountered in Lincolnshire.

Bob and his men had several run-ins with these poachers when often violence would be used as they attempted to escape. On one particular night Bob, his under-keepers and the police were called to a poaching incident in one of his woods. They surrounded the wood in question before Bob and some of his helpers went in to look for the poachers.

Bob was fairly certain of the area of the wood in which they were hiding and, as he approached some thick rhodedendrum bushes, close by he heard the sound of a .22 rifle bolt being

cocked. He knew now he was dealing with some very desperate men and quietly led his men to another part of the wood. He was sure that had they confronted the man he would have used his gun and he decided, very wisely in my opinion, that a few lost pheasants were not worth the loss of life of himself or one of his under-keepers that may have happened if they had confronted the poacher and he had started firing indiscriminately.

I heard this story from Bob when he returned to Rothwell as head-keeper. He also confessed he was pleased to be back in Lincolnshire where poaching was considered to be more of a nocturnal sport than the potentially more deadly business which went on around the Nottinghamshire pits.

THE THRILL OF THE CHASE

Dave, a local part-time keeper, was feeding a small flight pond on the Lincolnshire Marsh when he began to suspect that someone was shooting his flighting ducks on the nights he wasn't there.

He and his mate, Mick, decided to lay in wait one night and parked their van behind a hedge some distance from the pond. They made their way along the hedge and crept down close to the pond before settling down to wait.

As dusk approached two mallard flew around a settled on the shallow water. Almost immediately there was a bang from a small-bore gun and as they moved towards the pond, they saw two shadowy figures in the half-light. They shouted and gave chase and managed to catch one of them.

The poacher turned out to be one of the four lads our head-keeper ErnieTurner had found taking their ferrets 'for a walk' all those years earlier. The second one out-ran them but Dave and Mick were pleased to have caught one of them, who had been carrying a fine drake mallard.

They locked him in the back of their van and then went to a nearby cottage to call the police. While they were away the man tried desperately to kick his way out but the van's door locks held and he was eventually taken to Louth police station where he was charged with poaching.

As part-time keepers working on the edge of Louth, Dave and Mick experienced quite a few problems with poachers.

Both men were full-time farm workers and they would feed

their birds either early in the morning or in the evening in the dark winter months. This had its advantages because it meant they were often about well after dark.

One night they found a car parked at a local beauty spot and, when they shone their torch inside, they noticed some specks of dried blood on one of the seats. On closer examination, they spotted a few pheasant feathers in the foot well in front of the passenger seat. Clearly, it could belong to a poacher and they decided to hide close to a nearby gate until the owner returned.

Some time later they heard the catch on the gate click and they knew they were in business and grabbed the man as he shut the gate behind them. When they shone a torch in his face they recognised him immediately as a man who was a well-known local long distance runner. They also noticed he was carrying an air rifle and had a number of dead pheasants strung about him.

There was a scuffle and the air rifle and the pheasants fell to the ground. Dave warned Mick that if they let go of the man he would be off and, with his abilities as a runner, they would never be able to catch him again. So he held the man fast by his throat while Mick bent down to examine the pheasants.

SEVEN - EIGHT - NINE..!

"Six... seven... eight... nine. You greedy b*****d! Ten... eleven... twelve... thirteen. You greedy b****r! Fourteen... fifteen... sixteen... seventeen... eighteen..."

Between them they managed to bundle the man, the birds and the air rifle into their van and took him to Louth police station where he was charged with poaching. When he appeared before the local magistrates he was convicted and fined £36.

Dave and Mick loved shooting and field sports of all kind and regarded any poacher or trespasser as someone who was depriving them of their sporting rights with their employer. They knew that once he had had his few days shooting, they could treat the shoot as their own or could fish or ferret throughout the season.

Not all the poachers they came across were as fleet of foot as the runner. One day they came across a man ferreting their best rabbit warren. They went to speak to him and asked his name, to which he replied 'Mr Brown'. They asked him to take up his nets and leave the field and he agreed and started to do as they requested. As he did so, Dave noticed he was walking with a pronounced limp and, when they asked him about it, he told them he had a wooden foot.

It was only then that the penny dropped. Some time earlier a well-known local poacher had had a serious motor-cycle accident and had lost part of his leg.

Dave asked him again what his name was and he was told once again it was Mr Brown. "You're not called Brown," spluttered Dave, "you're Peg-leg ....!"

Knowing the game was up, the man quietly picked up his nets and limped out of the field, a warning not to come back ringing in his ears.

During one of their early-morning forays, Dave dropped Mick off to do some feeding before he headed for another wood some distance away.

When he came out of the wood he noticed a man with a greyhound coursing on a nearby stretch of grass land. He gave chase and the man headed towards where Mick was working, Dave yelling for Mick to stop the man.

By the time Mick realised what was going on, the man had gone past him and he joined Dave in the chase. By now they were heading for the River Lud, alongside which were some trout ponds.

Finally, Mick cornered the man between the Lud and the ponds. But the man was determined to escape and simply waded across the pond, the water coming up to his waist, before climbing out on the far side and disappearing over a wall into the nearby streets of Louth.

The poor chap was never to be seen coursing hares again, the chase by Mick and Dave perhaps having scared him even more than it did the hares he had been after earlier. It makes me wonder that if the government does ban coursing, will it also ban chasing coursers?

THE DEADLY DUO

Jim Kirk was born in 1931, one of his family's eleven children who were raised in the Bardney-Wragby area of Lincolnshire.

From a very early age he was interested in the countryside and, like many other Lincolnshire teenagers, turned to poaching during the Second World War.

Jim got a few birds to help feed his large family and his efforts helped put meat on the table during that difficult time. He also learned thoroughly the trade of poacher and, by the time he had reach 20, he was a highly-skilled night poacher.

As he roamed the countryside he became more and more skilled at hiding and avoiding detection. He was helped by being extremely fit, something his exploits demanded. It is no mean feat to be able to carry a large number of dead pheasants and negotiate several fields on your way home in the early hours of the morning.

His reputation grew and, as it did, he was asked by a number of his friends if they could join him on his night-time forays. Some lasted just one trip, others a few more, but no one had the bottle needed to match him as a night-poacher. That is until he teamed up with his brother-in-law, Jeff Speed.

Jim later related the story about how he once took his own brother with him. He was very jumpy throughout the expedition and was so scared when a pheasant fluttered up from the stubble as they walked home that he accidentally fired the gun he was carrying. He certainly didn't pass Jim's test.

By the time Jim and Jeff joined forces, large-scale pheasant rearing and releasing was taking place in Lincolnshire. Their target was not always dead game for, in the August-September period, they would often raid release sites as they had plenty of customers for poult pheasants as well as for shot birds later in the year.

Jim Kirk worked for the County Council Highways Department and travelled the roads of the area every day. He used his time profitably, keeping a close eye on game bird activity while he went about his work, which included inspecting roads for repair.

If he noticed a release pen being stocked up with young poults and he had a customer lined up, he and Jeff Speed would raid the pen during the next couple of nights, grabbing the young birds on the ground and putting them in sacks before throwing the filled sacks over the wire enclosure. They then carried them to the nearest roadside until they could be picked up by an accomplice, usually a woman, driving a car.

As long distances were involved they each carried two sacks so far, put them down and then went back and collected two more apiece, repeating the procedure until all the sacks were delivered in relays to the picking-up point. Doing it this way they could move up to 200 pheasants a night, with each sack holding up to 25 birds.

They targeted release pen sites on the Hainton, Rothwell, Stainton-le-Vale, Stenigot, Somersby, Fulletby, Tealby and Nocton Estates in addition to other isolated woods. From the routine practices on these estates they knew that when the birds had matured there would still be a large number of roosting birds in the vicinity of the pens.

Jim Kirk later maintained he was frightened of no one, having done some amateur boxing in his youth, but he much preferred to go about his business undetected. To help him do this he had a radio tuned to the police wave bands so he had advance warning of where the poultry check sites had been set up. He also knew a few policemen who were partial to a brace of pheasants so sometimes he was assured his back was covered by a friendly nod or wink from one of the officers in the area.

He and Jeff Speed used a variety of weapons for their poaching, mainly smooth-bore garden guns with .22 ammunition. They also used a variety of cars, most of which were driven by their wives.

They would be dropped off at a point close to where a gun had been hidden earlier in a drain, wrapped in a polythene sack to keep it dry, and they would have a pre-arranged pick-up point which may have been several miles away. Once on foot, they made for the nearest release site and plundered as many pheasants as they could.

Jeff Kirk said one of his best nights was when he shot 40 pheasants out of a single thorn bush, quite an achievement even for a highly-skilled poacher. Both men carried small torches and always tried to hit a bird in the neck. Then, having picked up the birds and carried them to the nearest pick-up point, they would move on to the next wood and start all over again.

When they had finished their night's poaching they would hide their guns in some convenient place, again wrapping them in waterproof sacks, and go to their pick-up point. If the car had not been seen or stopped and the radio indicated the coast was clear, they would back-track to their collecting point, load the birds into the car, collect their gun and make good their escape. They could always command a premium for the birds they poached because they were always shot in the neck.

Once when poaching the estate at Tealby, they disturbed another gang of poachers who fled, leaving behind their bag of pheasants and a very nice smooth-bore poaching gun, which Jim Kirk made sure he soon put to good use.

On another occasion, they were poaching at Somersby when he came to the edge of a wood carrying some birds for a pick-up later in the night.

He explained: "When I looked out into the field I could see a figure out there. I laid low for a while but the figure didn't move but I was unhappy about its presence as I thought it could be a keeper playing it very carefully.

"After a while the figure still hadn't moved but I couldn't stay there any longer so, for my peace of mind, I slowly walked straight towards it. I thought that whoever it was, they might feel threatened and retreat. However, it still didn't move and as I got nearer I found it was a newly located scarecrow. Certainly earlier in the day it hadn't been there. Having said 'good evening' to Mr Scarecrow, we then carried on our night's work."

On another occasion the pair were at Fulletby and quite close to the release pen they discovered a parked car. "I thought it had been parked there to deter us but I wasn't sure so Jeff stayed in the hedge with the guns while I walked round

the other side of the wood and approached the car from the trees. I crept up to it and, being fairly certain it was empty, pulled the door open. My assumption, was right, it had been left there to stop us entering the wood.

"I went back across the field to where Jeff was and, now fully armed, we went back and poached the wood. As we passed the car on the way out I pumped five rounds through the wind screen as a reminder to the keepers not to sit waiting for us in a vehicle!"

Some time later they visited Fulletby and had already shot 49 pheasants when they were caught by the gamekeeper, George Dawson, who was accompanied by the police and some local farmers. The two men were actually inside the release pen when they were caught, the poachers being heavily outnumbered.

Later their pick-up car was also stopped and, as a result of the events that night, Kirk, Speed and Speed's wife appeared before Horncastle magistrates on 17th February, 1977.

Mrs Speed was fined £20 for aiding and abetting the taking of game while Kirk and Speed were each fined £245 for poaching. When the magistrates' clerk asked Kirk how he wanted to pay, he calmly took out his wallet and, carefully opening it, replied famously: "Cash sir, I had a good night last night!"

Back in 1977 £245 was a very big fine but Jim Kirk's reaction that day showed just what contempt he had for the law and it represented the only time he and Jeff Speed were actually charged with an offence. Both, however, were to have several narrow escapes. By relying on field-craft and their widespread knowledge of their surroundings and their quarry, they always got away.

DECOYED

A well-know wildfowler, Jack Smith, once had serious trouble with poachers.

He owned two very good flight ponds on the Lincolnshire Marshes near Conisholme. Both ponds were very shallow and much favoured by teal when the moon and tide were favourable.

At both ponds 40-gallon oil drums had been sunk into the banks and were surrounded by reeds and they were used as hides. Very cosy they were, too, with no cold feet on this shoot for the wildfowler.

Jack favoured moonlit flighting and on the night in question had taken his friend Dick to man the other tub. They set a raft of decoys out and retired to the tubs to await the arrival of the ducks, which they expected to flight in the moonlight when the tide washed them from their roost on the sea shore some

four miles away.

Everything was quiet for about an hour when, suddenly, four shots were fired across the pond at the raft of decoys. As none of the decoys moved or took flight, the two men who had carefully crept up to the bank of the pond realised what they had done and made a quick dash for it.

Their hasty departure was accompanied by some choice words from Jack and, as they were silhouetted as they crossed the river bank on the way back to their car, they must have known what the decoys felt like a few minutes earlier.

TINNED PHEASANT

Not long after I started game keeping we had a shoot at Acthorpe and, at the end of the day, Ken Barker, one of the under-keepers, was not in the game larder as we counted the bag.

I asked if anyone knew where he was and the head-keeper, Ernie Turner, told me that at the end of the final drive while he was picking up, his dog had been attracted to a 40-gallon tin barrel used for storing feed for the birds.

When Ken took the lid off to look inside he found the bodies of a couple of pheasants. They had obviously been dropped there by one of our beaters, who planned to return later that night to collect them.

The bin was just inside one of our woods and Ken had been asked to remain there to keep watch and see who came back for the pheasants. During the day we had had some workers from a nearby farm beating for us and any one of these could have been responsible.

Ken kept the bin in sight until just after dark when he could make out someone on a cycle coming down the road. When he got near the bin, the rider dismounted, propped his bike against a tree and walked into the woods towards the bin.

Watching from behind a nearby tree, Ken saw the man go to the bin, lift the lid and reach down to remove the pheasants. At that point Ken stepped out and put his hand on the man's shoulder and asked him what he was doing. As you would expect, the man almost jumped out of his skin with fright.

It took him several minutes, in fact, to get over the shock and Ken, who had no transport with him, had to wait until the man had recovered sufficiently before saying he would walk with him the mile across the fields to the head-keeper's.

The man, however, didn't want to leave his cycle behind and offered to ride round by road to the house, a distance of some two miles. Ken said: 'Fair enough, but I'm walking beside you."

This they did and some time later the man was shown into the house where the guns from the day's shoot were still gathered. He was paraded in front of Mr Nickerson who gave him a severe dressing down before listening to the man's story. He had a large family to support and had only intended to take the pheasants to provide food for them.

Mr Nickerson listened carefully to what he was being told and, understanding the man's plight, decided to take no further action, putting it down to a certain amount of misguided necessity.

ONE FOR THE ROAD

Another encounter I had with poachers happened a few years later when the neighbouring estate at Cadeby was being troubled.

We had developed a call-out system under which we helped one another when poachers were active and on this particular occasion I received a call from Cadeby's head-keeper, Bill Tennant.

He told me two men had been seen near a wood which had been poached over the previous few weeks. I later met his under-keeper, Chris Palmer, and was told the suspected poachers had gone into the Granby public house on the A16 Grimsby-Louth road at North Thoresby. We went there and waited in the car park to see what happened.

Before long a van arrived and we could see it was a woman driving. She went into the pub and soon came out with the two men we were interested in and they drove off on the main road towards Louth. We followed at a discreet distance.

The police had been alerted by the estate and we were now joined by a police vehicle. We pulled over and gave the police officer the registration number of the van we had seen and he followed them with us behind, later finding the van parked in a lay-by with only the woman driver in it. We suspected she had dropped the two men off and, while the policeman questioned the woman, we retraced our journey to a point where a dyke crossed the main road.

It was getting quite dark but, after we searched one side of the road, I crossed to the other and found both men crouching in the dyke. I shouted across to Chris, who by now was checking further along the dyke and, as I did so, one of the men tried to climb out but I pushed him back in on top of his mate..

By now Chris had arrived and we allowed both men to climb out of the dyke. They denied knowing anything about poaching and claimed instead they had been searching the dyke for something which had fallen off their van.

I asked them where their van was and they told me it was 'down the road'. At that point they decided to make a run for it. Chris followed them on foot while I went back to get the Land Rover. We deliberately let them stay ahead of us and you can imagine the shock they had when they arrived panting in the lay-by only to find the police waiting for them.

Both men and the woman were arrested and taken to Louth police station. In the meantime, Chris and I went back to search the dyke and eventually we found a bag containing 16 pheasants and a .410 shotgun.

We took these to the police station and the men later admitted the gun and pheasants were theirs. One of the men later explained they had gone out with 17 cartridges and, during their night's poaching, they had missed just once.

These poachers worked out of Grimsby and used different tactics from the Lincoln gang. They went into the woods very early in the evening so, by the time we went out at about nine o'clock on previous occasions, they had already done their work and were probably back in Grimsby with their haul by then.

OUR DAILY ... PHEASANT

Throughout the history of rural England the role of the sporting parson had occasionally been mentioned but this particular story relates to a not-so-sporting member of the clergy.

Some considerable time ago the parish of Fotherby had a vicar whose actions certainly deterred one of his dwindling congregation to think seriously about listening to his sermons again.

Like many rural parishes, Fotherby was suffering from a decline in congregations. Right across Lincolnshire churches were recording record-low attendance and many faced closure.

Fotherby was no different and you could count on one hand the number of people in the village who were regular attenders. One of them was an elderly lady who lived close to the vicarage. She was a delightful lady and loved to feed the birds which were regular visitors to her garden.

Imagine her delight when the song birds and sparrows were joined by a splendid cock pheasant which would come every day to feed from the seed and tit-bits she put out for her visitors.

All was well until one day when she heard the unmistakable sound of a shotgun being fired near by. She looked out of the window and, to her horror, saw the vicar, gun in hand, picking up her beloved cock pheasant.

Perhaps God has his own game laws because I don't think that particular vicar had a license to shoot game and I suspect he didn't have a shotgun certificate either. In any event the death of this particular pheasant resulted in him losing his most stalwart parishioner.

TROUBLE WITH TRAVELLERS

In the middle of June in 1985 I had a call one Sunday afternoon to say several people were in a field on our boundary with some greyhounds.

I rang for some help from the police and then went to the field together with Steve Nicholson and Chris Palmer, a keeper from the neighbouring estate. On this particular day our local police sergeant was off duty but heard the call for assistance and turned up on his own motor-cycle.

When we arrived at the field, two vehicles were zig-zagging across a pea field with about ten or eleven men and children walking in all directions while several greyhounds were running around.

They saw us arrive and we waited at the entrance while they gradually came towards us, collecting their dogs as they did so. Then most of them climbed into one of the vans they had been using.

When they finally arrived we asked them what on earth they had been doing. They told us they had been exercising their dogs and didn't think they were doing any harm. They then all started talking amongst themselves in what we believed was Gaelic. We knew we had a party of Irish travellers in the area and had seen them parked not far from our shoot.

We explained to them that it was not acceptable to drive vehicles across growing crops — the peas were in flower at the time — upon which one of them turned to another man and said in a broad Irish accent: "I told you it wasn't grass. Grass doesn't have little white flowers."

Another of the men, who was dressed in a bright blue pin-stripped suit and fawn shoes, came up to us and said he had had nothing to do with any coursing. He was just visiting from Southampton, he explained.

The police sergeant said it would be best to ask them to move on. If we pressed charges the likelihood would be that by the following day they would have all left the county.

We took his advice and they agreed and drove off. They must have driven about five miles in a square for an hour later they were seen about two fields away on an adjoining farm doing exactly the same thing.

This same field was to be the scene of another encounter a few years later. It was early October and we had driven the field, which by then was stubble, for partridge and had moved on to the valley below. We then walked from the valley to get to our van, which we used to move the beaters about, and then drove along the lane at the side of the stubble.

While we were in the valley a hire van had arrived full of men and lurchers and when we drove past we could see three men with dogs out in the stubble while the others were still in the van, which was parked on the road alongside the field.

Several of the beaters offered to tackle the men in the field with their lurchers, but I told them to stay put while I had a word with the driver of the van.

When I got there the driver said he and his mates had come from Manchester. I told him we were in the middle of a day's partridge shooting and would they please go somewhere else. He listened to what I had to say, nodded his agreement and called the men in from the field.

It was clear the men with the dogs had done no harm and proved to be quiet amiable. This was one of those situations where a quiet word proved to be a far better way of dealing with what could have turned out to be a tricky situation.

Some years later the same area was again being visited by

men with dogs, this time from one of the neighbouring villages.

Several farm workers saw them from time to time and whenever they were asked what they were doing they were invariably abusive and threatening. As no one ever saw a hare actually being caught, they were more of a nuisance than anything else.

However, one day a young man from one of the local farms saw them in the field with their car parked on the roadside. He stopped his own vehicle and took photographs of their car. He then drove along another track until he was close to where the three men and their dogs where walking and he took some more photographs of them.

He then went to the police station in Louth and told them what he had seen and said he would bring in the photographs within a few days to support his complaint about the men.

I don't know what, if any, action the police took but the young man's quick thinking seemed to pay off for we were never troubled by these men and their dogs again.

THE PERFECT WITNESS

This particular story illustrates how times have changed and how the attitude of the police had altered in relation to law breaking in the countryside.

On February 2nd, 1990 at 10 o'clock at night I saw some lights shining in the field opposite my house. There seemed to be at least two lights and, before I could do anything more about it, my good mates Dave Harness and Mick Mamwell rang to say they had seen the same lights. Like me, they believed someone must be coursing and would come to help. I then called Louth police station, explained what was happening and arranged to meet their officers at the end of our lane.

When I went down the lane I found a yellow van parked on the roadside. It clearly belonged to whoever was in the field. By now I had been joined by Dave and Mick and we watched the field for a few minutes before three police officers turned up.

We all stood by the van and watched the lights in the field for a good 10 minutes and we could see the lights moving around and occasionally they picked up the greyhounds chasing rabbits and hares.

After a while I asked the police if they wanted the men out of the field as it did not seem likely they were yet ready to return to their van. They thought it was a good idea and, as I had a portable light with me, Dave, Mick and I drove across the field illuminating sections as we went along. The minute we went into the field the men stopped lamping.

In the middle of the field was a small copse and initially we could see no one. Then our light picked them up hiding on the edge of the tree and we could make out eyes of a pair of greyhounds. We believed there were three men but, when we drove up to them, we could see only two and the dogs. We told them to go back to their van where the police were waiting to talk to them and we escorted them back across the field.

On their arrival the police questioned them and made notes of their names and addresses and asked them where the third man was. They denied all knowledge of anyone else being involved but a little while later a third man was seen coming from behind a nearby hedge. While we had watched the men lamping, we had definitely seen the dogs catch something but now all three were empty handed. We were certain that when they spotted us, the third man had made himself scarce and had hidden their kill, intending to pick it up later.

The three poachers, their dogs and van were taken to Louth police station to be interviewed while Dave, Mick and I went in search of the spot where we believed they had probably hidden their catch. We hadn't been searching long before we found a bag conveniently hidden at the roadside for an easy pick-up. It contained the bodies of five rabbits and two hares and all were still warm and clearly were freshly killed.

We took the bag back into Louth and handed it in to the police as clear evidence of what the men had been up to. A couple of days later I saw one of the police officers concerned and asked them what had happened to the poachers.

He told me they had been questioned and then allowed to leave and, before doing so, were given back their lights, batteries and, remarkably, the rabbits and hares and told they would be hearing from the police at a later date.

I didn't hear anything more until 26th July, some six months later, when a policeman arrived at my house and said he had come about the poaching incident back in February. He said his sergeant needed a statement from me before making a decision whether to prosecute the poachers or not.

I was absolutely astonished at this long delay in the proceedings and then being asked to make a statement when three police officers had witnessed exactly the same as I had. But I duly made my statement but, as far as I know, the case never went to court.

This was the last poaching incident I was involved in where I called for police assistance. As far as I am concerned, if the police are going to condone such behaviour then there is little wonder there has been such a moral decline in our towns and villages.

Around the time of this incident the level of rural policing seemed to decline and it became very rare to see a policeman, or even a police car, in country areas. As the police moved into the computer age it seemed as if more and more of their dwindling numbers were tied to desks and not out on the beat any more.

Today, if we come across a poaching incident we have to ring the East Lincolnshire police headquarters at Skegness, a good 30 miles away. If a police officer is then required he or she would be sent from Louth, about five miles away.

Compare this with the situation years ago when all we did was contact our local village policeman. Now we have to explain the problem to someone on the other end of a telephone who is sitting 30 miles away and in all probability does not know the area you are calling from. Then, by the time assistance does arrive, the wrong-doers are miles away. There's little wonder that I can't remember the last local poaching conviction.

AN ARMED RESPONSE

As a contrast to the previous story, I have come across a couple of incidents recently where the police response has been, shall we say, slightly more than some people would have considered appropriate.

On one occasion, a keeper saw a couple of boys enter one of his woods with an air rifle. He telephoned for police assistance and, during the course of the call, was asked if the boys were armed. He said he thought he had seen them carrying an air rifle to which the reply was that the police would be sending an armed response unit to deal with the incident!

The next thing the keeper knew he was virtually besieged by armed police while a helicopter equipped with all the latest gadgetry hovered overhead.

This surely was a case of bureaucracy gone mad: a team of highly-trained armed police officers and a helicopter, all to catch two young lads out in a wood with an air rifle. Just imagine what the cost of all this was to the tax payer in Lincolnshire.

The second incident took place on the Laughton Estate, near Gainsborough when head-keeper Bill Richardson was alerted to a white car seen patrolling his lanes. A local farmer had spotted the car with a gun poking out of one of the windows and an occupant taking pot-shots at his pheasants.

Bill went in pursuit of the car but, when the driver spotted him, it took off at a fast speed in the direction of nearby Gainsborough. He followed the car for some distance before finally losing it on one of the housing estates in the town but he had managed to make a note of the registration number and went into the local police station to report the incident.

The officer on duty said he would log the incident and try to trace the owner of the car through the registration unit, but he decided not to mention the presence of firearms because that would trigger the involvement of the armed response unit.

Later that evening the same car was seen again on the Laughton Estate. This time the poachers were lamping the roadside fields and shooting at game trapped in the beam of their light.

Bill and his son, Jim, went out once again after the car, this time using two vehicles. In the ensuing chase Jim's vehicle and the car containing the poachers collided. The accident happened outside Bill's house and the five occupants of the car, angered at their night's poaching being disturbed, got out and

chased Jim into the house where he joined his family.

The group outside then turned very nasty, hammering violently on the door and trying to force their way in through a window. Inside, the occupants rang the police and asked for urgent assistance, explaining in detail what had happened. The police said they would attend as soon as possible but warned it could be a while before their men could get there.

Those inside the house spent a very frightening hour waiting for the arrival of the police while the poachers continued their siege, shouting threats and banging on doors and windows.

Finally, at about 10pm, a police armed response unit arrived, again supported by the police helicopter, which was equipped with a powerful search-light. Some of the poachers ran into a nearby wood but, with the aid of the helicopter and its light, they were soon rounded up and in no time at all the five poachers, three men and two women, were face down on Bill's lawn before being searched and then hauled off to Gainsborough police station.

The following day a policeman arrived to inspect the two vehicles that had been involved in the collision, the poachers' car and Bill's old van, one used on the estate to feed birds. As it was only used on the estate it was not taxed for use on the roads and the fact that no tax disc was displayed was the first thing the police officer picked up on.

He recorded all the details of the incident and its location, at a spot where a private road meets the public road, and said everything would be included in his report to his sergeant.

Bill heard nothing more for several days and finally called in at the police station to find out what action was going to be taken. He was told the accident was regarded as a 'fifty-fifty' incident and that no charges would be brought, either against the poachers or against Bill's son!

The result was that the poachers, despite their actions and the violent threats they had made, had got off scot-free. Bill's only consolation was that, after their encounter with his van, at least they wouldn't be doing any more poaching in that white car.

THE POACHER WHO SAILED
AWAY

When I was a young keeper I spent many week-ends on Reeds Island, a small island in the Humber close to the village of South Ferriby.

I used to accompany my boss, Mr Sam Nickerson, who visited the island regularly with other shooters in pursuit of wildfowl, my role being to act as his assistant, loading his guns and working my dog to pick up the birds he killed.

Often I would accompany other keepers in this work and we would spend the entire week-end on the island and, as we usually only flighted the wildfowl at dawn and dusk, we had some spare time during the daylight hours.

One Sunday, after going out for a dawn flight and returning to the house on the island for a good breakfast, the peace of Reed's Island was shattered by distant gunfire. Now on this particular Sunday I was the only visiting keeper so I went with the warden, who lived on the island and looked after it, to the western end from where the sound of shooting had come from.

It was about eleven o'clock in the morning as we approached the western tip of the island and we could see a figure on top of the sea wall. As the island was flat and we were hardly inconspicuous on a small grey Ferguson tractor towing a sledge so it was no surprise that we were quickly spotted.

By now we could make out it was a man with a shotgun and, on seeing us, he took to his heels and disappeared over the bank towards the river. When we reached the top of the bank we saw him climbing on board a cabin cruiser and starting its engine.

As the boat chugged away we shouted to him. He smiled at us and shouted back: "Hard luck — you can't cross that!", pointing at 10 yards of fast-flowing River Humber between the bank and the cruiser, and off he went on his merry way, the only time I can recall a poacher escaping from me by boat!

Reed's Island was owned by Mr Sam's brother, Mr Joe (later Sir Joe) Nickerson and used for grazing livestock and wild-fowling in the winter.

Over the period of his ownership, Mr Joe (as he was known to us all) thought it would be a good idea to introduce different animals onto the island and, over the years, these included brown hares, blue hares and fallow deer.

The two species of hare were the first to be moved onto the island and, under the stewardship of the first keeper there (he also acted as shepherd), they flourished and, during our visits, it was not unusual to see a few hares running across the closely-cropped pastures.

Some years later another keeper took over the dual role on Reed's Island and it was only after that when we began to notice a gradual decline in the hare population. After a couple of years questions were being asked about the disappearing hares and he came up with the story that only recently, while returning to the island by boat, he had spotted a hare swimming in the opposite direction. Hares can swim but this explanation seemed most unlikely.

This particular shepherd had several dogs he used during shooting along with collies which helped him look after the sheep on the island. During one of my visits I went with him to feed his dogs when he carried a couple of buckets. One contained boiled meat and the other biscuits and he mixed the contents of both into the bowls for the dogs.

I couldn't help but notice that the meat looked suspiciously like hare and I asked him about it. He was honest with me. "Mr Joe wonders what's happened to the hares. Well, that meat there is from the last one on the island." He said it had taken a lot of catching but he had finally got the last one.

Some keeper!

ON-THE-SPOT JUSTICE

As the level of rural policing has declined so Lincolnshire has become the target for all-night raids of gangs of poachers from outside the county.

They come with four-wheel drive vehicles, powerful lamps and lurchers and ride all over farm land in the middle of the night, either sending their dogs after their quarry, which include hares foxes, badgers and deer, or shooting them with high-powered rifles. As only the hares and deer can be used for food the driving force seems to be sport rather than financial gain.

One of the incidents I came across involved a farm near Wragby where, early one morning, the farmer was told there was a four-wheel drive vehicle and some men in one of his fields. Now this particular field had a few wet patches in it where underground springs had come to the surface and, when he arrived, he found that it was in one of these spots that the vehicle had become bogged down and its occupants were desperately trying to get it out.

The farmer immediately knew what had happened and realised that, no matter what the men did, they were not going to free their vehicle. He went over to them to ask what they were doing in the middle of his field and the ring-leader explained they had lost one of their dogs and had been simply driving across the field to try to locate it.

However, the farmer knew this wasn't true and told him: "I think we both know what you're doing here and the way I see it is this: I can leave you and your vehicle here and go and ring the police or you can give me £100 and I'll send my man over with the big tractor and pull your vehicle out and we'll forget you were ever here."

The poacher thought for a while before answering. "We don't want the police." Then he opened his wallet and counted out £100. Their vehicle was duly towed free and the men left. This all happened long before the police introduced on-the-spot fines for motoring offences!

The second incident took place in 1999 and clearly illustrates how dangerous lawlessness in the countryside can be.

A gang of five poachers from South Yorkshire went onto a farm at Stewton, near Louth in search of badgers and foxes and were equipped with a Land Rover, lurchers and lamps.

After catching one fox they stopped for a smoke and it was at this point that the farmer, Mr Ward, became aware he had intruders on his land. He considered ringing the police but thought then would take too long to reach his farm so he took his .22 rifle from its case and decided to confront the poachers himself with the gun loaded for his own protection.

When he approached the Land Rover he became seriously concerned for his own safety and fired a few shots over the vehicle with the idea of frightening the men from his land. Unbeknown to Mr Ward, one of the poachers was standing on the passenger seat of the vehicle with his head out of the sun-roof and, as he raised his hand to adjust his night vision goggles, one of the bullets from the farmer's gun struck his hand, shattering several bones.

The poachers then beat a hasty retreat, taking their injured colleague to the local hospital for treatment to his hand. They also decided to report the matter to the police. As a result, the farmer was arrested and taken to Louth police station where he was charged with unlawful wounding.

When the matter went to court Mr Ward was cleared of all the charges. However, he surrendered his firearms certificate and was told he could not reapply for it until the following year. In the meantime, no charges were brought against the men who went onto his land and, once again, the poachers got off scot-free.

Although the farmer was mistaken in taking the law into his own hands in this way, he felt that, faced with five men in a Land Rover equipped with powerful lights, he was in personal danger and only fired the shots as a warning. Additionally, it must be remembered that if the poachers had respected the rights of the farmer and had not been trespassing on his land, the incident wouldn't have happened in the first place.

Incidentally, I understand the man who was injured went on to make a full recovery only to be later arrested for poaching. He obviously learned nothing from his experience that night in a field at Stewton.

A POACHER NAMED STEVE

The stories that follow were related to me by a man I will name only as Steve. He lives in the west of Lincolnshire and his poaching escapades are certainly worth relating. Steve was a man who poached mainly to provide food for his family and concentrated mainly on running dogs although he had been known to shoot the odd roosting pheasant or two.

The first story he told me, in fact, related to pheasants and happened at Sandbeck Park on the Nottinghamshire border.

Steve's mate had been on Sandbeck with his lurcher only to run into a keeper who shot his dog. Naturally, Steve's pal was greatly upset by this so he, Steve and another lad decided to get their own back on the keeper.

Although Steve was only a small-time, one-for-the-pot poacher, he and his two pals raided the Sandbeck woods and, in four hours undetected poaching, left with 116 pheasants in three sacks. This was a large haul for just three men and if the keeper had only been a bit more diplomatic when he found Steve's pal with his dog, then perhaps his wood would never have been raided.

A CASE OF MISTAKEN IDENTITY

The next story Steve told me also involved roosting birds. When he was thirteen he had a pal named Charlie whose parents didn't always see eye to eye.

One night Charlie's parents had a right set-to and it ended

with his mother threatening her husband with a .410 shotgun. This, naturally, incensed Charlie's dad and he grabbed the gun from his wife, bent the barrel over his knee and threw the weapon into an out-house.

Some time later the two lads came across the gun and decided they would have a go at straightening the barrel out. They put it into a vice and, after a lot of work, decided the barrel was nearly straight.

It was time to try it out. They took the gun to a local sand pit and set up a stone on the bank as a target. When the first shot was fired, a large cloud of dust flew up on the bank some way to the left of the stone they were aiming at. However, by mental adjustment the lads managed to bag a few rabbits that day with their 'nearly-straight' shotgun. As their parents knew nothing about them having the gun, the decided to hide the weapon at Steve's house.

In the meantime Steve had heard how poachers liked to shoot roosting pheasants and one night decided to have a go himself at bagging a few of these tasty game birds.

He went into a wood he knew was used by pheasants and, after creeping about for a while, found some birds roosting in the trees. Using the same mental adjustment for the gun, he managed to hit all thirteen birds he had seen.

They all fell into some dense undergrowth and, in the darkness, he could not recover any of them so he decided to go back the following morning. This he did and, after a bit of searching, managed to recover all thirteen birds. However, much to his dismay, they were all bantams, something he hadn't been able to make out in the darkness of the previous night

THE LAST CAST

Steve tells the next story in his own words.

"As a young lad I used to fish in a small river on the far side of the Hodsock Estate, near Blyth, in Nottinghamshire. It was not exactly teeming with fish but there was enough to keep a young lad interested.

"The problem with the place was getting to it without being seen or by having to make a detour of some four miles. The keeper on the estate at the time was a Scotsman called Andy, who was very keen on his work. However, I was also very skilled indeed when it came to sneaking about without being seen and it was this skill which saved my bacon on one particular occasion.

"I was fishing the river one afternoon in early October. It had been a lovely day and I had caught a few fish and, as the sun was just about to set, I thought I would have one last cast by an oak tree. It had always been a good spot in the past but it was difficult to approach because of the open nature of the banks.

"My trick was to take to the nearby woodland and then come straight to the bank. However, when I rounded the last bramble patch there was Andy not ten feet away, facing the river with his gun under his arm. I'd almost pushed him into the river with the end of my rod!

" I reversed carefully for about thirty yards without daring to turn before quietly making my way back home. But he didn't see me and I lived to fish the river another day."

WET FEET

At Steve grew older he became more interested in running dogs and this became his stock-in-trade. Here is another story, again in his own words, of his exploits in this field.

"Once while I was out lamping on my own, I had caught four rabbits and decided to switch the lamp off for a few minutes. As I did so another lamp was shone into the field I was in from the adjacent roadway.

"I crept along the hedge-row and, after I had covered about 100 yards, the beam of the mystery lamp again scanned the field from the roadside.

"I came to the end of the hedge where it bordered a stream, which was about fifteen feet across and some eighteen inches deep. I waded across and then stood among some nearby trees to watch, only to see the lamp come on again. This time the beam was very close but it stopped just before it got to me.

"Suddenly, a dog shot out and caught a rabbit trapped in the beam. It was only then that I realised I had got very wet indeed for the sake of another poacher whose dog I recognised in that instant. I later told him I had been watching him at work that night but I never let on about my dip!"

AN UNNERVING EXPERIENCE

Steve told me about the night he and a pal were out on a lamping trip. They were crossing an open stretch of land covered in dead bracken in a place known as Crossley Wood.

"My companion, Lee, was about five or six yards in front of me when I suddenly felt rather than saw something push past me. It was so clear I even felt pressure on my left foot. At the same time my dog pulled hard right backwards.

"I immediately spun round and banged the lamp on and scanned around the patch, which must have covered around two acres. But I could see nothing, nor could Lee.

"Every time I went to that area of Crossley Wood in the future I always thought back to that night and, as I did, the short hairs on the back of my neck stood up."

THE STRANGE COMPANION

This was not Steve's only experience of coming across something he could not explain while out on his nocturnal activities.

"I was out looking for a bird or two and was walking very carefully along a ride through a wood. As I looked along the ride I could see the figure of someone silhouetted against the lights of a nearby village. Whoever it was was moving from side to side, first from one side then to the other.

"I slipped off the ride in case I was seen but kept watching carefully for whoever was there. At first I thought it might be another poacher. However, there were birds all around me and I heard no shooting and whoever my strange companion was, he couldn't have been more than fifty yards away.

"I looked again and he had gone so I laid low for about half an hour before I got on with what I had come for, shooting about a dozen pheasants before leaving the wood. I never did know who or what was with me that night."

THE COACH PARTY

Poaching can have its funny side and Steve told me about another experience he had which still makes him laugh today.

"While out on a lamping trip to a golf course alongside the A57 we were catching rather a lot of rabbits when I suddenly

heard a vehicle approaching. I switched the lamp off as this was the usual thing to do to avoid detection, particularly when lamping near a road.

"Vehicles had passed us before but this particular one stopped, so we kept quiet and watched from our vantage point, which was some twenty yards from the roadside hedge.

"After a few seconds about half a dozen young women came through the hedge, all laughing and giggling like tipsy women do. Then they all proceeded to answer a call of nature right in front of us.

"After a few seconds we realised what was happening and, as it was quite clear they were no threat to us, I switched on the lamp and flashed (no pun intended!) across to where they were all squatting. Well, you should have seen their reaction! It was sheer chaos and that does not even begin to describe what happened. You just couldn't believe how quickly a party of drunken women can get their drawers up and get back on their coach!

"It both made and ruined our evening because, after that, we could not really get serious about anything for the rest of the night."

SPINNING A YARN

Steve and his pal Kevin had been on a lamping foray with their greyhounds and, to allay any suspicion, had left their van some way from the scene of their activities.

After catching several rabbits and hares it was arranged that Kevin would head across country with the dogs and their catch while Steve went back to the van and would drive round to pick him up at the end of the lane. If anything went wrong, he was to head back and they would meet in the local rugby club's car park.

Steve headed in the opposite direction to return to the van. After driving off he realised he was being followed by what looked very much like a police car.

Kevin, in the meantime, had remained alert and when he saw two pairs of headlights coming towards him in the lane, he knew something was wrong and decided to lay low.

Steve also twigged what had happened and, reaching the end of the lane, carefully turned round and drove back towards the police car. When the vehicles were close the police driver signalled Steve to stop and, when he did, he was asked what he was doing out in the countryside at three o'clock in the morning.

The few minutes between spotting the police patrol and being stopped had given Steve time to get his story straight. He explained that earlier in the day he had brought his greyhound for a walk in the lane and it had run off. As the bitch was worth some £3,000 he had spent the previous few hours looking for it.

He must have been convincing for the policeman seemed to accept it straight away and showed some concern for the missing bitch, carefully noting down a full description of the missing greyhound. Craftily, Steve had given him a detailed description of a brindle and white bitch he had at home, his thinking being that if the police made any further inquiries, he could always say he had found the dog when he went home.

The policeman was thorough, however, and insisted in searching Steve's van. Needless to say, it was empty apart from some straw in the back which, Steve explained, had been bedding used by the 'lost' dog.

While all this was going on, Kevin had made his way to the rugby club and was waiting for his friend. The police officer told Steve he would circulate a description of the missing dog. Steve thanked him and, as they parted, said he would drive down to the rugby club just to have another look round before he called it a night and went home ...

THE DESERTED FARM

Another story Steve told me related to the time when he used to poach rabbits from a spot near Barnby Moor, near Retford, where he was always assured of a good catch, either with greyhounds or ferrets.

"The farmstead belonging to the land had been deserted for years and the present farmer lived miles away so we never got any bother there.

"When myximatosis spread through the rabbit population, it took a heavy toll in this area and we sort of forgot about this bit of land for seven or eight years. Then, one night my mate Kev suggested we pay it a visit with the lamp, something we had never done there before.

"Don't ask me why, but I agreed to go that night without making my regular day-time reccy. I had always made it my rule that I never went anywhere without first sussing out all the options before a night-time visit.

"Nevertheless, off we went and all was going well, we were catching steadily and there seemed to be plenty of rabbits on the land.

"But as we got near the farmstead we noticed a couple of cars parked outside. It was too late for them to belong to any workmen but, as we were used to poaching loads of game within yards of farm buildings and even keepers' cottages, we didn't see much danger.

"This time, however, it was different. The very next rabbit the dog chased went round the corner of a wall and out of sight. All of a sudden all hell broke loose. The dog was screaming, other dogs were barking, lights were going on all over the place.

"I ran round the corner and grabbed Kev's dog, which had run into some electric netting at full speed and had become entangled. As I tried to free it, I was constantly getting electric shocks. It took me some time to get the dog clear but finally we both made good our escape.

"We found out later that in the intervening years the farmstead had been sold and converted into five homes, which all goes to prove my point that you should never poach anywhere until everything has been checked out.

"If I had taken the trouble to check that land we would still have gone lamping that night but you can be assured we would have avoided the pantomime we found ourselves engaged in."

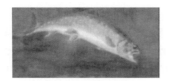

THE SOLE SURVIVOR

Brian had a passion for fly-fishing and once he had acquired some land of his own he thought it would be a good idea to dig a trout lake so he could fish to his heart's content.

After some time and a great deal of work he had his 'lake' some 15 metres wide by about 80metres in length. It ran near his new bungalow and, once it was ready, he stocked it with some 200 rainbow trout. It was his great joy to stand alongside his own lake and throw his trout a handful of feed and watch the water boil as they all came up for a tasty morsel.

As the trout grew to maturity he and his friends got out their rods and caught a few, releasing most back into the water. By controlling their catching in this way, Brian and his friends had many hours of pleasure from the well-stocked trout lake. That was until one weekend when Brian and his family were away on a short holiday.

When the family returned to their bungalow Brian walked down to his beloved lake and cast a handful of feed on the water. Then he waited, and waited. Finally, a single trout rose to the surface to take some of the food.

Brian was puzzled by this response and waited for some time at the side of the lake. But no more trout were to be seen. He went back into the bungalow and asked his young daughter if she had been out before him to feed the trout, but she said she hadn't. He returned to the lake later and tried again. But again only one trout came to the surface to take the food.

In the past he had never been bothered by vermin like mink or by otters, heron or cormorants so, as he stood there, he was at a loss to know what had happened to his trout.

He decided to walk around his lake to look for clues and when he reached the farthest point from his home he found tell-tale footprints. It was clear from what he found that at least two people had come onto his land while he was away and stolen virtually all his fish. They had probably stunned the fish using a powerful electric current or had simply strung a net across the lake and then, by walking along both sides and dragging the net along, simply trapped the trout.

Although the lake was some 15 feet deep they had managed to net all the trout except, it seems, just a solitary fish which had evaded the poachers.

This true story indicates just what devastation can be caused by one poaching gang trespassing on land where they had no right to be. Brian lost all but one of his fish and his freedom to pursue his hobby on his own land by the actions of this gang of trout poachers.

This incident took place in the mid-1990s and since then Brian has been reluctant to restock his lake because he fears that once the news got out the same fate would befall him again. Because of the actions of these poachers he has lost his fishing and the enjoyment he derived from feeding his trout.

JIM'S SHARP REPLY

The story that follows happened around the time of the millennium and clearly illustrates the lawless situation which now prevails in much of the Lincolnshire countryside.

Jim, a farmer in the south of the county, had his attention drawn to a vehicle and two men with greyhounds who had been seen on his land one Sunday morning.

Now Jim was then around 60 years old and of average build while the two men he found on his land were in their early twenties, one of average size while the other weighed around 18 stone.

As Jim approached the two men, who he clearly thought were poachers, he was immediately told to f*** off but Jim wasn't a man that easily deterred. He was a plucky character and had farmed this land for some 40 years and was determined that no ruffian was going to turn him out of his own field.

He realised straight away how precarious his position was as the larger of the two men started coming towards him in a menacing position, particularly as he had nothing to defend himself with. Then he remembered he had a ten-inch cauliflower cutting knife behind the seat of his four-wheel drive vehicle. Jim grabbed the knife but it proved no immediate deterrent as the poacher made it perfectly clear he wasn't frightened by it.

The poacher then took a swing at Jim but, in doing so, he cut his wrist on the blade of the knife. He immediately retreated, nursing the gash and Jim could see he had blood dripping through his fingers.

By now the second man had armed himself with a half brick and was threatening to use it on the farmer. Jim, however, was determined to get these men off his land and walked towards the man, warning him that he would face dire consequences if he tried to use the brick.

The poacher looked across at his mate, he was holding his wrist and had blood running down his hand, and realised that Jim meant business. He dropped the brick and the men collected their dogs and drove off.

Jim went home and immediately telephoned the police to report the incident. Two hours later an officer arrived and asked Jim what had happened. He explained how aggressive the men were and said that when one of them tried to take a swing at him he pricked his hand on a knife Jim had picked up to defend himself with.

The policeman asked if the men had left the land and when Jim said they had he replied: "Good, that's all I want to know." He then bid Jim good day.

THE RED-FACED KEEPER

In the 1960s there was a shepherd at Rothwell named Jack who had probably poached the odd hare or pheasant in his time and liked to pull the leg of the head-keeper, Jacobs, and tell him he would never be able to catch him.

Jacobs became obsessed with this and instructed his under-keepers to keep a sharp eye on Jack because he just knew he was poaching.

Jack, of course, loved it and this went on for some time and Jack would always make sure either Jacobs himself or one of his under-keepers was within ear-shot when he boasted about how tasty the bird was he had enjoyed for his Sunday lunch.

One evening in the gathering dusk as the beaters left the yard, Jack made a point of bidding the head-keeper 'good night' and, as he walked away, two or three tail feathers from a cock pheasant could be seen protruding from the lunch bag slung across his back.

As he went round the corner Jacobs looked up and saw the feathers and immediately shouted and ran after Jack. He finally caught up with him and believed at long last he had trapped the wily shepherd.

Now Jacobs was a man with a ruddy complexion and he seemed to positively glow as he demanded that Jack empty his lunch bag. "I've been watching you for a long time and now I've finally caught you!" he gleefully declared.

Jack looked totally innocent. "You can look in here, the bag's empty," he said with a grin on his face. When Jacobs opened the bag all that fell out were three long tail feathers from a cock pheasant.

Head-keeper Jacobs' face was now redder than ever as he realised he had been tricked. And he never did catch Jack.

THE BEST LINCOLNSHIRE POACHER OF THEM ALL

All the stories I have related so far have been about poaching incidents I have come across during my life, through my own experiences, by word of mouth or when the poachers concerned were caught and prosecuted.

Over the years there have been several who could claim the title of The True Lincolnshire Poacher. I certainly can think of three men who were real contenders.

The first was Kenzie Thorpe who certainly had a justified claim but whose career was spoiled by the number of times he was caught. It wasn't surprising, really, for his favoured poaching weapon was a 12-bore shotgun, a noisier gun you couldn't wish to use. In his favour, however, was his impressive bag of game, but I believe there were two men who were both better poachers than Kenzie and were certainly more discreet.

The first of these was Geoff Cummings, the leader of the Lincoln gang whose exploits are told at some length in this book and the man who wrote to me with that detailed account of the gang's activities.

During my career I have been at the other end of Geoff's poaching operations and, believe you me, he was very professional and in a career spanning over half a century, he was only caught a few times and prosecuted on just three occasions.

I now come to the man I believe was the best poacher I ever

encountered. His name was Norman Elvin and he has already been mentioned briefly. He was one of the two brothers among a gang of six poachers who raided the Worlaby shoot and killed 140 pheasants in a single night.

Norman was a man I knew all my life. He was born the son of a gamekeeper in 1931 and moved with his family to South Elkington, near Louth, in the early 1940s where his father was gamekeeper in Welton Vale.

As the son of a keeper, he quickly picked up the ways of the country and soon became proficient with a gun, snare, net or trap. He started poaching for pocket money while he was still at school, although I'm not sure whether he poached his father's pheasants or preferred to go elsewhere for his sport.

Norman Elvin

By the time I started keepering Norman already had quite a reputation as a poacher, although because he had never been caught and because he told only a select few about his activities, information was hard to come by.

When I was a young lad just starting out as a keeper, one story about Norman came to light and made me realise that if he ever came on our estate, we would have a hard job catching him.

The story came from Norman's brother. Like many keeper's cottages, their home was in the middle of a wood and it seemed that on this occasion Norman had a bet with his brother that he knew the wood so well he could walk blind-fold through it and out to the village shop about a mile away. On the day in question Norman's brother personally tied the blind-fold and he and some of his pals watched as Norman carefully made his way through the wood and down to the village shop to claim his reward.

This was just one of the little snippets we used to build up a picture of a man blessed with rare poaching talents.

Between the ages of fifteen and thirty Norman poached a lot of birds, usually using a .22 rifle fitted with a silencer and it was only some years after the event that I heard of him and the others taking 140 birds in one night from Worlaby.

Although very little was revealed, it shows the scale of the poaching in which Norman was involved during this time when he was never caught, let alone convicted.

As his family grew older Norman retired from poaching and, perhaps inevitably, became a part-time gamekeeper. The skills he had learned and used so effectively as a poacher stood him in good stead in his keepering job and many a fox has met its downfall because Norman knew how to remain undetected.

From an early age he had a hatred of foxes, probably because his own free-range chickens had suffered at their hands from time to time. Norman usually had the last laugh and many was the morning he was seen walking home with a dead fox slung over his shoulder.

He also possessed the rare skill of being able to call rabbits from their burrows and was also very adept at calling stoats. Not surprisingly, Norman was an expert rabbit catcher and spent many happy hours out ferreting.

A couple of weeks before the great Countryside March in London in 1998 he was in a hedge bottom rabbiting when the local organiser of the Lincolnshire contingent spotted him and pulled up in his car. He got out and asked Norman what he would be doing a week on Sunday. "Ferreting," came the reply.

The march organiser asked if he was interested in joining the group on the march in London but Norman told him he didn't think it would do any good and, as far as he was concerned, as long as ferreting wasn't banned he couldn't care less about fox hunting. The organiser was left in no doubt that here, at least, was one countryman he couldn't count on for support.

Norman's detestation of foxes extended to all vermin. He was very good at calling carrion crows and had some very choice words for these birds, which were one of the many threats to young pheasants. If he could call a carrion crow he was one of the happiest men alive and it was almost as if, after poaching pheasants for much of his life, he was now trying to redress the balance in favour of the pheasants by exterminating as many of their natural enemies as possible.

Norman lived on the edge of our shooting land in the early days but, as we acquired more land, his home became closer to the centre of our shoot. By this time we were firm friends and he became an enormous help in assisting us in stopping predators. Norman rarely kept an eye on the time and was happy to stay out for hours at a stretch if it meant catching another fox or stoat.

One amusing story I recall occurred one snowy morning. A fox had visited his yard and he followed the foot prints to a false fox-earth situated not too far away from his home. He noted the fox had entered the earth and, because he had a full-time job at the time on a building site, decided to leave things as they were, planning to return after work with his gun to wait for the fox coming out again at dusk.

Some hours later I visited the same earth, coming at it from the opposite direction, and noted the same footprints along with a set which unmistakably belonged to Norman (in later years he had a real moocher's gait and left very distinctive prints).

I popped my terrier in one end of the earth and soon the fox came out the other and I shot it. This all took about a minute and then I retraced my steps, buried the fox and went on my way.

That evening as darkness fell, Norman arrived home, grabbed his gun and went straight to the false earth. He selected a bush to stand behind, believing he wouldn't have long to wait before the fox came out. It would be silhouetted against the snow and he believed he would be able to shoot it quite easily.

It was a bitterly cold night and, after standing there for two hours, he finally decided something had some awry with his plans and that the fox wasn't going to show itself.

When he approached the earth he quickly saw why. He saw my footprints and, a little further along, came to the spot where I had shot and buried the fox. What words he uttered under his breath at that moment I'm not sure, but knowing Norman as I did I'm sure he wasn't too upset because another fox had clearly met its end.

In later years he helped on our shoots at North Ormsby and Acthorpe and really enjoyed the experience of being among shooting people.

It was while helping on our game cart in November, 1998 that Norman collapsed and died at the relatively young age of 67. His friends all said that was the way Norman would have wished to go, among shooting people and with a pheasant in his hand.

We have two mementoes left to us by Norman. One was a game rack he designed and made and the other is a wooden seat placed near the best pheasant drive in Welton Vale by his family. It bears the most fitting epitaph, which reads:

In memory of NORMAN HENRY ELVIN 1931-1998
Who learned his trade in these woods
as a poacher and then a gamekeeper.

Having looked carefully through the exploits of my three candidates, I would place Mackenzie Thorpe third as, although he was probably dealing with wild game, some of his stories are somewhat open to argument and his kills-to-conviction ratio was not impressive.

Geoff Cummings, on the other hand, would come second because although highly professional, much of his poaching was done amongst reared pheasants which were much easier to kill than wild pheasants. In addition, he usually worked with at least one accomplice.

Although Geoff's kills-to-conviction ratio was quite high it could never match that of Norman Elvin who, to my knowledge, was never caught. He poached both wild and reared pheasants and was quite clearly the true Lincolnshire Poacher.

Norman usually operated on his own, hence we never heard too much about his activities and, unfortunately, he took most of his stories with him to his grave.

They say a good poacher is as crafty as a fox and Norman certainly outsmarted a lot of foxes in his time.

Derek Mills and his wife Gill, pictured after he had received an award from the Country Landowners' Association in 1998 marking his 40 years association with the North Ormsby estate.

POACHERS' TALES

The stories recounted in this book are as true and as accurate as is possible. By its very nature, poaching is a difficult subject to record but, during my search for different stories across this wonderful county of ours, I have endeavoured to stay within the bounds of fact as near as I possibly can.

I regret that I have obviously not managed to include every account of poaching I have come across because, like all those fine fishing stories, sometimes where truth finishes fantasy begins. Time had also dimmed the memories of both poachers and gamekeepers and I have had to be careful where exaggeration has taken over. In addition, many of those involved have passed over to that other great hunting ground and have taken their wonderful stories with them.

I apologise to all the living poachers who are not mentioned. No doubt, they will have bigger and better bags to record in future. What I have done is endeavoured to present a cross section of the poaching which has gone on across the fields and through the woods of Lincolnshire during the last 60 years.

It may not always have been a shiny night, but it was certainly their delight come the season of the year!

Derek Mills,
Louth
March, 2001

ACKNOWLEDGEMENTS

I should like to thank the following for their help in the preparation of this book:

Louth Library, Gordon Ward, Maurice Elvin, Steve Colclough, Richard Elvin, Pop Ireland, Jim Measures, Dave Harness, Jeff Cummings, Andrew Kay, Jim Kirk and others who have helped to complete the stories.

There are also two others who have died: Tom Parker and Alan Holmes.